Programming PIC
Microcontrollers with PicBasic

Programming PIC Microcontrollers using PicBasic

by Chuck Hellebuyck

Newnes

An imprint of Elsevier

Amsterdam   Boston   Heidelberg   London   New York   Oxford
Paris   San Diego   San Francisco   Singapore   Sydney   Tokyo

Newnes is an imprint of Elsevier

Recognizing the importance of preserving what has been written, Elsevier prints its books on acid-free paper whenever possible.

Library of Congress Cataloging-in-Publication Data

ISBN-13: 978-1-58995-001-6
ISBN-10: 1-58995-001-1

British Library Cataloguing-in-Publication Data

A catalogue record for this book is available from the British Library.

The publisher offers special discounts on bulk orders of this book.

For information, please contact:

Manager of Special Sales
Elsevier
200 Wheeler Road
Burlington, MA 01803
Tel: 781-313-4700
Fax: 781-313-4882

For information on all Newnes publications available, contact our World Wide Web home page at: http://www.newnespress.com

10 9 8 7 6 5 4

Printed in the United States of America

Dedication

This book is dedicated to my wife Erin and my children Chris, Connor, and Brittany.

This book would never have happened without your support.

Contents

Introduction

Electronics has been my hobby and profession for over 25 years. I started as a young child building kits from Radio Shack and projects described in electronics magazines and books. When microprocessors were first developed, I was fascinated with them. I was a bit too young to really understand how they worked, but I could see they would replace the batches of discrete integrated circuits (ICs) my previous electronic projects depended on. I soon discovered microprocessors required many more tools and resources (like money) than I could afford. This made it difficult to build a home lab for micro-based designing so I never got involved during all the early years of microprocessor development.

I went on to earn a bachelor's degree in electrical engineering and made electronics my profession. Although I had learned how to program and work with some of the best microprocessor tools, I still didn't see the opportunity to build a home lab for microprocessor development without spending a bunch of money.

Then I discovered the Microchip PIC family of microcontrollers. They were inexpensive, easy to purchase through various sources, and development tools were inexpensive. I bought a PIC programmer and started playing with electronics as a hobby again. Although I developed some interesting projects using Microchip assembly code, I really longed for a simple form of programming like the BASIC language because I didn't have a lot of spare time.

A company named Parallax began advertising a small PIC-based computer module called the "Basic Stamp" that could be programmed in a form of BASIC. I bought one and I started playing with it. It was easy to use, and I had a lot of fun with it. But it had memory limitations and was a bit expensive to make permanent designs with. I had spent a lot of time developing gadgets and really wanted to turn a couple of my ideas into products I could market.

I thought about developing my own Basic compiler for the Parallax computer module that would allow me to program a PIC directly. Then I saw an advertisement for a new product from microEngineering Labs called the PicBasic compiler. It could convert a program written for the Parallax module into the code format required to program a PIC. It used the same commands as the Parallax module along with a few more. I purchased one immediately and began designing in PicBasic.

I found it to be a simple but very powerful compiler. I could develop complex projects in a few days rather than weeks or months with assembly language. I designed a few products and began to market them through my website at www.elproducts.com. I also decided to write an article for *Nuts and Volts* magazine about the Microchip PICs and fortunately got it published in July 1998. I was then approached about writing a book on PICs. I never thought of myself as an author but I saw it as an opportunity to share my knowledge about PICs and PicBasic with those who might enjoy this stuff as much as I do.

As I wrote, many things got in the way and this book took far longer to write than I had originally expected. But the delay allowed this Basic programming method to become more popular. New compilers from other companies, new programming accessories and hardware began to show up all over the place. The PICs and the PicBasic compilers improved as well.

As it evolved and my own experience increased, I tried to capture as much as possible in this book but still keep it at the entry level. One result of my increasing experience was to modify the original outline to include a chapter on robotics. Robotics has become very popular during the time I wrote this book, and I believe it's because there were more people like me who were using all the new affordable yet powerful microcontroller tools to develop robots in their home labs.

Using Basic to program microcontrollers began to be called embedded Basic programming and recently I've seen job postings for PicBasic programmers. It's become harder to find people who are trained at programming in assembly code, with so many electronic development companies switched to the C language. I believe embedded Basic will be the next wave of programming for small module high-volume designs since it's so much easier to write and almost as efficient as C.

I hope you find this book informative and challenging, not to mention enjoyable. Everything in here was learned the hard way—by trial and error. Microchip has some great components and the PicBasic compiler makes it easy for everyone to become an embedded Basic designer. You can visit my website for more info on some of the latest embedded Basic products. If you have any questions, I can be reached via email.

Chuck Hellebuyck
Electronic Products
www.elproducts.com
chuck@elproducts.com

CHAPTER 1

Getting Familiar with PICs and PicBasic

The PIC (*Programmable Interface Controller*) line of microcontrollers was originally developed by the semiconductor division of General Instruments Inc. The first PICs were a major improvement over existing microcontroller because they were a programmable, high output current, input/output controller built around a RISC (*Reduced Instruction Set Code*) architecture. The first PICs ran efficiently at one instruction per internal clock cycle, and the clock cycle was derived from the oscillator divided by 4. Early PICs could run with a high oscillator frequency of 20 MHz. This made them relatively fast for an 8-bit microcontroller, but their main feature was 20 mA of source and sink current capability on each I/O (*Input/Output*) pin. Typical micros of the time were advertising high I/O currents of only 1 milliampere (mA) source and 1.6 mA sink.

General Instruments eventually sold its semiconductor division, along with the PIC manufacturing facility in Chandler, Arizona, to a venture capitalist group that formed what is now known as Microchip Technology. PICs quickly became the main components offered by the new company.

Initially the selections were small and none of them had common microcontroller features such as timer overflow or external interrupts. They also used a somewhat unusual banking arrangement for memory that still exists today in many of Microchip's parts. Despite these limitations, the PICs sold well and allowed Microchip to develop new components with new features including interrupts, onboard A/D (*Analog/Digital*) conversion, on-board comparators, and more.

Microchip's lineup soon included flash memory components as well as low-cost OTP (*One Time Programmable*) devices. These low-cost OTP devices set Microchip apart from their competitors. Other 8-bit micro companies offered OTP components, but they usually came at a high price premium relative to masked ROM (*Read Only Memory*) versions.

Masked ROM microcontrollers are fabricated by placing layers of semiconductor material on top of each other to form the transistors and other components. The proper arrangement makes the microcontroller operate according to the software. After a masked ROM is created, it cannot be changed. Even one software command change requires a new masked ROM. Microchip found a way to produce OTPs at only a small cost premium compared to masked ROM parts. This allowed designers to use OTPs in final designs because small changes could be made without stopping production or spending more money for a new masked ROM.

Microchip also made their PICs serially in-circuit programmable. This allowed a manufacturer to build up electronic modules with an unprogrammed PIC on-board and then program it right on the factory floor. That flexibility made Microchip popular with professionals as well as experimenters. Microchip has since grown to become the second largest producer of 8-bit microcontrollers. Microchip also expanded to become a leader in low-cost, long-life EEPROM (*Electrically Erasable Programmable ROM*) memory and other niche markets.

Microchip continues to develop new microcontrollers at a rapid pace with the devices falling into three main categories: 12-bit core, 14-bit core and 16-bit core program memory. All the parts have an 8-bit wide data bus that classifies them as 8-bit microcontrollers. No matter what your application, Microchip probably has a device that will work well with your design concept.

PIC Overview

This book focuses on programming PICs in the PicBasic language. The PicBasic compiler (*PBC*) is designed to work with the popular 14-bit core devices. The PicBasic Pro compiler (*PBPro*) works with the 14-bit core, 16-bit core, and the new 18CXXX components that don't have the page limiting memory all the other PICs have.

I cannot cover all the devices from Microchip in this chapter since the PIC family continues to grow. However, I want to give you a basic overview of the

Microchip microcontroller devices you will most likely be working with. Later in this book, I'll spend more space detailing some of the inner workings of the 14-bit core components. My intent is not to give you a summary of the Microchip data book, but instead to help you understand how to properly write programs to control a PIC.

I will mention assembly language from time to time because that is the programming language Microchip developed for PICs. Many professionals program in assembly and even Basic programmers should have some knowledge of assembly language. Don't let that scare you though; I'll show you how to use the PicBasic compiler so assembly language will be something you rarely use.

Consider this section to be the fundamentals—the stuff no programmer really likes but the stuff every programmer should know!

The PIC family can be broken up into three main groups, which are:

- 12-bit instruction core (16C5X, 12C5XX, 12CE5XX)
- 14-bit instruction core (16C55X,16C62X, 16C6X, 16C7X, 16C71X, 16C8X, 16F8X, 16F87X, 16F62X, 12C6XX, 16C9XX, 14C000)
- 16-bit instruction core (17C4X, 17C7XX, 18C2XX, 18C4XX)

All three groups share the same core set of RISC instructions, with additional instructions available on the 14- and 16-bit cores. This means that assembly code written for the 12-bit family can be easily upgraded to work on a 14- or 16-bit core device. This is one of the great advantages of the PIC.

Another feature is that all assembly language instructions (except `branch` and `goto` instructions) execute within one clock cycle (crystal frequency/4), which makes it easy to check the execution timing. That isn't the case with the PicBasic language, since it compiles higher-level commands into groups of assembly code.

Once you have compiled a PicBasic file, it creates an assembly file. If you understand assembly code, you could work with that file. Most users won't need that. It's only when doing advanced PicBasic programming that you may need this detail. After creating the assembly file, the PicBasic compiler will assemble it into the binary (`.hex`) file needed to program a PIC. That binary file is then used to actually program the PIC using a PIC programmer.

An abbreviated list of PIC devices and brief list of features are outlined in Table 1-1.

Table 1-1: Abbreviated list of PIC microcontrollers and their features.

Device	ROM Words	EEPROM Bytes	RAM Bytes	# I/O	A/D	Timers	Misc.
12 bit Core							
12C5XX	0.5K to 1K		25 to 41	6	none	1+ WDT	8 pin package
12CE5XX	0.5K to 1K	16	25 to 41	6	none	1+ WDT	8 pin package
16C5X	0.5K to 2K		25 to 73	12 to 20	none	1+ WDT	18 pin, 28 pin package
14 bit Core							
12C67X	1K to 2K		128	6	4	1+ WDT	8 pin package
12CE67X	1K to 2K	16	128	6	4	1+ WDT	8 pin package
16C55X	.5K to 2K		80 to 128	13		1+ WDT	18 pin package
16C6X	1K to 8K		36 to 368	13 to 33		3+ WDT	18 pin, 28 pin, 40 pin package
16C62X	.5K to 2K		80 to 128	13		1+ WDT	18 pin package
16C7X, 71X	.5K to 8K		36 to 368	13 to 33	4 to 8	3+ WDT	18 pin, 28 pin, 40 pin package
16F87X, 8X, 62X (FLASH)	.5K to 8K	64 to 256	36 to 368	13 to 33	0 to 8	3 + WDT	18 pin, 28 pin, 40 or 44 pin package
16F9XX	4K		176	52	0 to 5	3 + WDT	64 or 68 pin package, built in LCD driver
14000	4K		192	20		1+ WDT	28 pin package
16 bit Core							
17C74X	4K to 16K		232 to 454	33		4+ WDT	40 or 44 pin package
17C7XX	8k to 16K		678 to 902	50		4+ WDT	64 or 68 pin package

12-bit instruction core

This is the original core produced and is used in the most cost-effective parts available from Microchip. They use only 33 assembly language instructions. But because they only have a two-byte wide stack, these parts will not work with the PicBasic compiler. I've included them in Table 1-1 so you know they exist, but as prices of the 14-bit PICs have declined, the advantages of the 12-bit versions have faded.

14-bit instruction core

The 14-bit core parts are second-generation devices. Microchip added interrupts and other features, and a clever thing Microchip did was to keep the footprint or pin-out the same as for the 12-bit components. They also kept most of the 12-bit core assembly code instructions, allowing a direct upgrade from the 12-bit core parts to the 14-bit core parts without changing the circuit board or having to do a major software revision.

Because of the added features, the number of assembly instructions increases by two for a total of 35. Microchip actually added four instructions and replaced two 12-bit core assembly commands with special function registers. The two instructions replaced by a special function register are the TRIS (port direction) and OPTION (special function).

The four added instructions include two math function commands and two return commands. The two return commands include one return command for the interrupts and one for subroutine returns, which can be nested deeper on the 14-bit core because the stack increases to eight levels. This increased stack size is necessary to use the PicBasic compiler.

Table 1-1 lists the feature summaries for these parts. They also offer most of, if not all, the features any electronics hobbyist needs to develop microcontroller-based products.

16C55X

The 16C55X is pin-for-pin compatible with its 5X 12-bit core cousins, but with a major addition: interrupts. They also add one more I/O pin by sharing the TOCKI external clock pin (used for incrementing the 8-bit timer from an external source). The interrupts include the 12CXXX wake-up on state change interrupt along with a real interrupt pin for capturing an event. Also included is a timer overflow interrupt for the 8-bit timer. All the interrupts jump to a single redirection register, so your main interrupt routine will have to bit test the interrupt flags within the INTCON register. Your program can mask any and all interrupts through the INTCON register also. A final difference is the I/O characteristics increase to 25 mA sink and source.

16C62X

These devices are similar to the 16C55X group but add two on-board comparators to the package. The 62X components have 13 I/O and 0.5k, 1k, or 2k of 14-bit wide code space. They share all the features of the 14-bit core group including the interrupts. If you need comparators in your design then these could reduce your overall parts count.

A new device recently released by Microchip was the 16F628. It is a flash version of these components.

16C6X

These parts were part of the original 14-bit core group and consist of several devices with unique features. They start with the 16C61, which isn't much different from the 16C556 part, but the rest of the 16C6X group is very different. They add the following features to the devices previously mentioned: 2k, 4k, or 8k of code space for programs, 22 or 33 I/O, synchronous serial port (shared with I/O), one or two Capture/Compare/ PWM pins (shared with I/O,) and three timers (two 8-bit, one 16-bit).

The 16-bit timer is great for accurate timing requirements. It can run from its own crystal separate from the main clock source. It will even run during sleep mode, allowing time to increment while very little current is being consumed by the PIC. It has an overflow interrupt so you can wake up from sleep process the timer information and then sleep some more.

The synchronous serial port can be used to communicate with serial devices. It operates in two modes: 1) serial peripheral interface (SPI), or 2) inter-integrated circuit (I^2C).

These are very powerful components.

16C7X, 16C71X

These parts are identical to their 16C6X cousins with the addition of four, five, or eight channels of 8-bit on-board A/D conversion. For example, if your design uses a 16C62 and you need to add A/D, you can drop a 16C72 in its place. They are pin-for-pin compatible with each other. The A/D converters are shared with some of the Port A and Port E I/O pins, so its best to save these when doing a non-A/D design that may later need A/D. The 16C71X devices are upgraded versions of some 16C7X parts that add more RAM space.

16C67X

These parts are the 8-pin package versions of the 14-bit core group. They share the I/O the same way the 12CXXX 8-pin parts do to maintain one input only and five I/O. The amazing thing is that they also have four channels of A/D conversion that operate the same as the 16C7X devices (shared with the I/O). Code that was written to work with the 16C7X A/D will work on the 16C67X. They also have all the 14-bit core interrupts, and one 8-bit timer with timer overflow interrupt and built in oscillator option. They offer 0.5k and 1k of code space. This is a lot of microcontroller in a small package.

16C8X, 16F8X

If you're looking for a flash or EEPROM version of the PIC, this is the group. Originally Microchip only offered EEPROM versions (16C8X) but now have released them in flash (16F8X). They have all the features of the base 14-bit core group: interrupts, 13 I/O, one 8-bit timer, 0.5k or 1k of code space as EEPROM or flash and 36 or 68 bytes of RAM.

Unique to these devices is the 64 bytes of EEPROM data memory. This data will stay even when power is removed so it's great for storing calibration or variable data to be used when the program starts again. They are very handy for development because they can be programmed over and over again without ever leaving the circuit.

16F87X

This is one of the newest groups of devices from Microchip. They have flash program memory so they can be reprogrammed over and over again. They are built to be identical to the 16C7X family with some data memory and program memory updates. They offer 22 to 33 I/O, three timers and up to 8k of program memory. They have all the special functions the 16C6X and 16C7X parts have as mentioned earlier.

All the projects in this book will be built around the 16F876 because it is flash reprogrammable, has A/D, and has all the other PIC features. It also offers the option to build a *bootloader* inside. A bootloader allows you to program the part from a serial port without any special programmer circuitry.

16C9XX

This device shares many of the 16C63 and 16C73 features (three timers, interrupts, etc.) but adds another feature: on-board liquid crystal display (LCD) drive circuitry. It can drive up to 122 segments using four commons. The 16C924 also has five channels of A/D on-board, making this a great component for measuring analog signals and then displaying the results on an LCD.

With the 16-bit timer, it could display time for possible data-log applications and with the synchronous serial port any kind of external data storage or PC interface is possible. These devices seem to have it all except on-board EEPROM for nonvolatile memory storage.

14C000

This is a different numbering scheme and offers a different approach. It's a mixed-signal processor. It has a slope-type A/D, instead of sample and hold, and also has D/A (digital-to-analog) conversion capability. It shares the higher-end 14-bit core characteristics, including the three timers and such. These are unique devices when compared to the rest of the PIC family but share the same code.

16-bit instruction core

This is the high-end group from Microchip. They cannot be used with PBC. To program these in PicBasic, you will have to use PBPro. That is one of the advantages that PBPro offers and why it costs more than PBC.

The 16-bit core parts offer up to 33-MHz clock speed for a 121-nanosecond instruction time. They have the same 35 instructions as the 14-bit core plus 23 more instructions. The stack increases to 16 levels. 33 I/O is standard with two open-drain high-voltage (12 V) and high-current (60 mA) pins. They add another 16-bit timer for four total timers.

These parts can also operate as a microprocessor rather than a microcontroller by accessing the program to be executed from external memory. These are not the parts to start experimenting with until you've mastered the 12- or 14-bit core parts. If you're experienced with other microcontrollers, then you may be able to use them right away.

This book is really dedicated to the beginning PicBasic user so I won't spend more time on these parts. You should now have enough basic knowledge to understand what the different PICs are about. Now I'll discuss software as we lead into using PBC and PBPro.

Software for PICs

A microcontroller is nothing without software. To program PICs requires a binary file of coded ones and zeros. Microchip offers an assembly language for PICs and a free assembler to get you going. Assembly language can be tough for a beginner, though. It is easier for a beginner or hobbyist with limited time to use a higher-level language and a compiler to convert that higher-level language into an assembly language program.

PicBasic is a higher-level language that is easy for beginners, hobbyists and even professionals to use for simple code development and rapid prove-out of a concept. I recommend it and use PicBasic often. I also write in assembly and recommend everyone learn it at some point, but PicBasic is a great way to start and in most cases stick with. Since this book is about PICs and PicBasic, I'll just touch on assembly below and then dive into the guts of PicBasic.

Assembly Language

All microcontrollers run on simple binary codes. These codes are various arrangements of ones and zeros. Assembly language is a higher-level language to this binary code and Microchip PICs have their own set of assembly commands. These commands when combined as a program are assembled by a software program called an *assembler*. The assembler outputs a file in the binary command form the microcontroller uses. That binary file is the "ones and zeros" program that controls the PIC.

Microchip offers a free assembler for software writers to assemble their programs. The file produced by the assembler for PICs uses the Merged Intel Hex format or INHX8M and is given the `.hex` file suffix. This `.hex` file is what the PIC programmer tool uses to burn the program into the PIC's program memory.

Assembly commands, although easier to understand than binary code, can be difficult to understand and can take a beginner months of practice to get a program to work. That's why even higher-level languages such as PicBasic became popular. At some point, though, you'll need to do something with the PIC that PicBasic or any higher-level language won't do. That's when you may want to use assembly language.

Sometimes a single assembly language command can solve the problem. PicBasic fortunately has the capability to mix assembly code within the PicBasic program. In the chapters where I discuss the various PicBasic commands, I'll show you examples of using assembly code.

I've written hundreds of programs in PicBasic and never had to use assembly language but it helps to know it's there when you really need it.

PicBasic Compiler

Back in 1995, a company named Parallax incorporated developed a small computer module based on the PIC that could be programmed in a modified version of the BASIC software language.

Parallax Inc. had been producing programmers and emulators for the Microchip PICs but saw a potential to make PIC-based design easier for everyone. They knew that assembly language programming was difficult for the beginner and hobbyist so they decided to develop a form of the BASIC language called PBASIC. They developed the computer module around a PIC 16C56 device and called it the BASIC Stamp. The module used external EEPROM memory to store the program, and the PIC retrieved commands from that memory one at a time and executed them. This is known as *interpreted execution*, which the BASIC language is famous for. Although this isn't the fastest way to run a program, it became popular with many experimenters, electronic hobbyists, and even professional technical people. It offered a totally new approach to programming PICs that was simple and quick.

It wasn't long before some users were asking if working programs could be compiled into assembly language so a PIC could be directly programmed instead of the somewhat expensive PIC-based Basic Stamp computer modules. Micro Engineering Labs answered the call. They developed a PicBasic compiler, or PBC, that would take a working PBASIC program and convert it into the INHX8M format required to program a PIC. They added more commands to increase the capabilities of PicBasic. It really made PIC-based development easy.

The compiler works with all the 14-bit core parts previously mentioned and when compiled a program will run about 15 times faster than the same program running on the Parallax module. Because the code is compiled rather than being directly written in assembly, it isn't as efficient as an assembly language program— but it can be close. The true advantage is reduced software development time. Programs that may take weeks or months to write in assembly can be written in days or weeks in PicBasic. For the professional, this offers quick concept "prove-out" or even rapid production. For the hobbyist or experimenter it offers quick project development and a shorter software learning curve.

I have found some limitations with PBC but can usually work around them with better program structure or occasional assembly language inserts. That was the case until the PicBasic Pro (PBPro) compiler was introduced. It offered so many features that I found I never had to add assembly code to my programs at all. It also could compile programs much more efficiently than the PBC.

These two different but related versions of the PicBasic compiler will be covered in this book, the standard lower-cost PBC version and the PBPro professional version.

I'll try to be consistent and call the professional version of compiler "PBPro" and the standard version will be called "PBC." This should make it easier to understand.

PBPro and PBC share the same basic code structure, but the PBPro version offers many added features and is really designed to be independent of the Parallax module coding limitations.

In Chapters 2 and 3, I'll give a brief overview of the PBC and PBPro commands, respectively. In later chapters, I'll show you examples of both versions at work in projects you can build yourself. Both versions include a manual and this book is not intended to be a substitute for those manuals. This book is intended to be a complimentary resource for making PICs, PBC, and PBPro easier to understand and use. The PicBasic language is really easy to learn and somewhat intuitive but the examples and explanations in this book should leave you ready to program any concept you have in mind. It's only limited by your imagination.

CHAPTER 2

PicBasic Compiler (PBC)

Programming microcontrollers in BASIC may seem old fashioned or limited in capabilities. After all, the BASIC language has been around a long time. It was so easy to learn that kids could program with it. The first Apple computers, Commodore computers, and Radio Shack TRS-80 computers all came with BASIC as their programming language. The BASIC language is what helped Microsoft's founders get started in business. So how could such an old language still be useful today? For all the reasons it was successful in the early days: the simplicity of the language.

Almost anybody can read a BASIC program and understand a few lines even if they have never programmed before. Microcontroller development, on the other hand, is not that easy. You need at least some knowledge of electronics. You also need some knowledge of algebra. And you need some knowledge of structuring a software program.

Building simple kits can help you pick up electronics knowledge. Algebra is something we all should have learned in school. But how do you simplify learning structured software development? By using an easy-to-understand language like BASIC. You don't need to know quantum physics to understand how a transistor works and you don't have to understand advanced calculus to understand basic algebra. So why should someone have to learn assembly language to program a microcontroller? Thanks to the PicBasic (PBC) compilers, programming Microchip's PICs can be easy for anyone.

In this chapter, I want to focus on just the PBC. It doesn't have all the commands and features found in the PBPro compiler, but that does not rule it out for many applications. PBC doesn't handle program spaces larger than 2k very well because of the PIC's inner structure, but a program of 2k is still quite large (and much larger than the Basic Stamp module). That 2k limit to PBC is something PBPro does not have and is why some people prefer the PBPro compiler instead. But I can tell you from my experience that the PBC is so efficient that I have written many very powerful programs that fit in a 1k 16F84A device. When you figure the PBPro compiler is almost two and a half times more expensive than the PBC, you just can't rule out the PBC. It's really a great compiler for the money.

In this chapter, I will cover each PBC command in some detail but won't replicate what you can find in the PBC manual. What I have done is expand upon the information in the PBC manual. I will also explain how to use the PBC compiler and give you a better understanding of the compiler's function. To understand how to use this compiler, though, it helps to know how it works. Let's start there.

How PBC Works

The guts of the PBC are a batch of short little assembly language programs written to do certain tasks. When the compiler is run, it groups those little programs together according to your PBC program structure.

If, for example, you want to turn an input/output (I/O) pin high so an LED will light, then you would issue the HIGH command in your PBC program. It's not that easy in the PIC, though. First you have to change the I/O pin to output mode. Then you have to set the bit within the port register that corresponds to that pin. This would take several commands in assembly code. A brief assembly code example to set bit 0 of Port B to a high state looks like this:

```
bsf   STATUS,RP0        ;Move to register bank 1
movlw  0FF              ;First make all pins of PORT B
movwf  TRISB            ;  high impedance inputs
bcf   STATUS,RP0        ;Move to register bank 0
movlw  01               ;Set bit 0 of PORT B
movwf   PORTB           ;  to high.
bsf   STATUS,RP0        ;Move to register bank 1
movlw 0FE                   ;Set PORT B pin 0 to output
movwf  PORTB               ;  and the rest of the pins to inputs
bcf   STATUS,RP0           ;Move back to bank 0
```

Although this probably isn't the most efficient way to do this in assembly language, it does show the several main steps required. The same function in PBC looks like this:

```
high  0                              'Set PORTB pin 0 to high
```

When the commands get more involved (such as serial communication) the assembly code file gets bigger but the equivalent PBC command takes just one line. This explains why higher-level languages are more efficient for the developer. The cost for that is the inefficiency of the assembly language the compiler creates. Some assembly language commands within the various compiler programs could be shared, but aren't because of the structure. The author of the compiler program tries to keep those inefficiencies to a minimum, but it's almost impossible to get rid of them all. That's the price we pay for quick, easy-to-follow program development. However, I've found the PBC to be quite efficient.

I do a lot of development with the 16F84 flash PIC that has only 1k of ROM space. When I've run out of space, simple modifications to my PBC program allowed some complex routines to fit. What really helps is the vast array of commands PBC offers. Serial RS232 type communication, lookup tables, and math functions are just some of the complex features PBC has reduced down to a single command. PBC includes the following list of commands:

ASM..ENDASM: Insert assembly language code section.

BRANCH: Computed GOTO (equivalent to ON..GOTO).

BUTTON: Debounce and auto-repeat input on specified pin.

CALL: Call assembly language subroutine.

EEPROM: Define initial contents of on-chip EEPROM.

END: Stop execution and enter low power mode.

FOR..NEXT: Repeatedly execute statement(s).

GOSUB: Call BASIC subroutine at specified label.

GOTO: Continue execution at specified label.

HIGH: Make pin output high.

I2CIN: Read bytes from I^2C device.

I2COUT: Send bytes to I^{2}C device.

IF..THEN: GOTO if specified condition is true.

INPUT: Make pin an input.

LET: Assign result of an expression to a variable.

LOOKDOWN: Search table for value.

LOOKUP: Fetch value from table.

LOW: Make pin output low.

NAP: Power down processor for short period of time.

OUTPUT: Make pin an output.

PAUSE: Delay (1millisecond, or msec, resolution).

PEEK: Read byte from register.

POKE: Write byte to register.

POT: Read potentiometer on specified pin.

PULSIN: Measure pulse width (10us resolution).

PULSOUT: Generate pulse (10us resolution).

PWM: Output pulse width modulated pulse train to pin.

RANDOM: Generate pseudo-random number.

READ: Read byte from on-chip EEPROM.

RETURN: Continue execution at statement following last executed GOSUB.

REVERSE: Make output pin an input or an input pin an output.

SERIN: Asynchronous serial input (8N1).

SEROUT: Asynchronous serial output (8N1).

SLEEP: Power down processor for a period of time (1 Sec resolution).

SOUND: Generate tone or white noise on specified pin.

TOGGLE: Make pin output and toggle state.

WRITE: Write byte to on-chip EEPROM.

Some of these commands will be used in every program you write, while others will only be used in specific applications. The list may seem extensive, but in time you'll find the commands are easy to remember and understand.

Variables, Memory, and I/O

The PBC was written to use the same basic structure as the Parallax BASIC Stamp module. The Stamp only allows eight I/O pins for program development. A standard 14-bit core PIC has at least 13 I/O pins available. The Stamp also has limited space for program memory and variables. Program memory is limited to 256 bytes, and RAM or variable space is limited to 13 bytes. The14-bit core PICs have an entry level of 512 bytes of ROM or program memory space with up to 8k available as upgrade parts. However, remember the PBC doesn't handle program space larger than 2k. The 14 bit core PICs also offer more I/O and more variable RAM.

To use the extra I/O and RAM, or variable memory in the PIC, and still maintain compatibility with the Basic Stamp module, the PBC just added additional commands and variable names. The added program memory space in the PIC didn't require any special commands. It naturally allows larger programs than the Stamp. This is a major advantage the PBC compiler has over the Basic Stamp.

For variables, the Stamp named each of its 13 predefined RAM locations bytes B0 through B13. Word variables are formed by combining two bytes. Of the 13 bytes, six byte pairs are used and are named W0 through W6. For example, W0 is the same space as B0 and B1 combined.

The first pair of bytes—B0, B1 that form W0—are also individual bit names. The least significant bit in B0 is labeled BIT0, the second bit BIT1, etc. This allows individual bits to act as flags without using up a whole byte.

The PBC takes advantage of the added RAM in various PICs. It adds more byte variable names along with added word names. Table 2-1 and Table 2-2 show the variable arrangement for the various 14-bit core PICs.

Table 2-1: Predefined PIC variables.

16C61,16C71,16C710,16F83,16C84	B0 - B21	W0 - W10
16C711,16F84	B0 - B51	W0 - W25
16C554,16C556,16C620, 16C621	B0 - B63	W0 - W31
16C558,16C622,16C62A,16C63, 16C64A , 16C65A,16C72,16C73A, 16C74A	B0 - B79	W0 - W39

Table 2-2: Predefined PIC variable alignment.

W0	B0 B1	Bit0, Bit1, ... Bit15
W1	B2 B3	None
W2	B4 B5	None
...	...	...
W39	B78 B79	None

The added I/O is handled by the special commands PEEK and POKE. Because the BASIC Stamp PIC-based module only offered eight I/O pins (which are actually the eight bits of the PORT B PIC register), all additional PIC I/O is accessed through direct manipulation of the PIC's port data and TRIS registers. This is a bit of a hassle but compatibility with the Parallax module forced that direction.

These PEEK and POKE commands really allow direct access to the PIC's internal registers similar to assembly language programming, but without leaving the PBC command structure. I'll talk about this in more detail in the POKE and PEEK command description, but note that any PBC commands that require a pin designator will only work on the eight PORT B I/O.

Program Operators

Symbols

Variables can be renamed using the SYMBOL statement. This allows PBC users to change the B0 format to anything they feel describes the variable more effectively. The format is simply:

```
Symbol    count = W1   ' W1 can now be referred to as count
```

Symbols must be at the top of the program. Symbols can also be used to set constants.

```
Symbol    Value = 10   ' Value can be used instead of 10
```

This is handy for having one location to change constants rather than changing them all the way through a program. When a symbol is used to define a constant, no RAM memory is used up. It's simply used as a compiler directive.

Comments

Comments within a PBC program can be formatted in two ways. The comments can be preceded by a single quote (') or the REM keyword.

```
HIGH   1     ' This would be the comment
LOW    1     REM This would also be a comment
```

Numeric Values

Numeric values can be specified in three ways: decimal, binary, and hexadecimal numbers. Decimal numbers are the default so nothing is required to tell PBC you mean decimal. Binary numbers must be preceded by the % symbol and hexadecimal numbers must be preceded by the $ symbol.

```
100          ' Decimal value 100
%01100100    ' Binary value for decimal 100
$64          ' Hexadecimal value for decimal 100
```

ASCII Values

ASCII characters must be placed within quotes. They are treated as the numeric ASCII value in all operations. Several ASCII characters together are treated as separate characters. These are mainly used when transmitting information with the SEROUT and SERIN commands.

```
"A"          ' Treated as ASCII value of decimal 65
"HELLO"      ' Treated as individual ASCII values for H,E,L,L and O
```

Line Labels

The PBC compiler doesn't allow or require line numbers for each program line. Sometimes a label is required to designate a location in the program for jumps and branches. This can be done with a label followed by a colon (:). Labels can be placed on a line by themselves or at the beginning of a command line. Labels are a necessary part of PBC programming. Labels are limited to a length of 32 characters and cannot start with a number.

```
Start:
                        ' Start program here

Finish: END             ' End program here
```

Math Operators

This is where the beginner and even the experienced user will appreciate the PBC compiler when compared to assembly language. PBC allows simple math instructions to be included right in the program. There's no need for advanced routines or bit manipulation; it's all done for you by the compiler. The list below shows the math operators.

   It's important to note that all math functions are performed strictly from left to right. This violates the typical math rules of parenthesis operations first, then multiplication, then division, etc. This can be confusing if you are doing complex items. It's best to break up functions to make it easier to follow. Breaking up the equations will not increase the memory usage in most cases.

+	Addition
–	Subtraction
*	Multiplication
**	Most significant bit (MSB) of multiplication
/	Division
//	Division remainder only
MIN	Limit result to minimum value defined
MAX	Limit result to maximum value defined

&	Bitwise AND
|	Bitwise OR
^	Bitwise XOR
&/	Bitwise AND NOT
| /	Bitwise OR NOT
^ /	Bitwise XOR NOT

All math is performed with 16-bit precision, which allows byte and word math. Multiplication is actually 16x16, resulting in 32-bit results:

```
W2 = W1 * W0    ' The lower 16 bits of the result are placed in W2

W2 = W1 ** W0   ' The upper 16 bits of the result are placed in W2
```

Division does the opposite:

```
W2 = W1 / W0    ' The numerator of the result is placed in W2

W2 = W1 / / W0  ' The remainder only is placed in W2
```

Math operators also include what I call "digital logic math." AND, OR, and exclusive OR can all be performed on variables. The opposite is also available: NAND, NOR and exclusive NOR. These commands are great for bit testing or bit manipulation without affecting the whole byte.

```
B4 = B2 & %11110000   ' Store the upper four bits of B2 in B4 and
                      ' ignore the lower four
```

MIN and MAX operators set limits for the variables. For example:

```
B1 = B1 + 1 MAX 128   ' B1 can increase to 128 but no larger

B1 = B1 -1 MIN 1      ' B1 can decrease to 1 but never 0
```

PBC Commands

Hopefully you now have a good idea of the program operators. They will become clearer when I show actual program examples in later chapters. Now we need to cover the guts of the PBC compiler, namely how the commands operate. To help explain the various command functions I've broken them down into separate groups.

I/O Control

This group contains some of the most commonly used commands. After all, most of the PIC's operation involves turning outputs high, low or reading a value.

HIGH *pin*

This command sets a specific bit in the PIC PORTB data register to high and then makes that pin an output. The *pin* value designates which PORTB PIC bit to set high. *Pin* must be a number from 0 to 7.

Example:

```
HIGH   1    'Set PORTB bit 1 high and make it an output. (PIC pin
            '7 on 16F84)
```

LOW *pin*

This command sets a specific bit in the PIC PORTB data register to low and then makes that pin an output. The *pin* value designates which PORTB PIC bit to set low. *Pin* must be a number from 0 to 7.

Example:

```
LOW   1     'Set PORTB bit 1 low and make it an output. (PIC pin 7
            'on 16F84)
```

INPUT *pin*

This makes a specific bit in the PIC PORTB data register an input or high-impedance pin ready to measure incoming signals.

Example:

```
INPUT 1     'Make PORTB bit 1 and input. (PIC pin 7 on 16F84)
```

OUTPUT *pin*

This makes a specific bit in the PIC PORTB data register an output. You must be careful to know what state the PORTB data register is in before issuing this command. As soon as you issue this command, the status of the bit in the data register (high or low) will instantly show up at the PIC pin.

Example:

```
OUTPUT 1    'Make PORTB bit 1 and output. (PIC pin 7 on 16F84)
```

TOGGLE *pin*

This command reverses the state of the port pin in the data register. If a port pin was high, it is changed to a low. If it was low, then it's changed to high. If the port pin was an input prior to this command, the port pin is made an output and then the state of that port pin in the data register is reversed.

Example:

```
TOGGLE 2    'Change state of PORTB bit 2. (PIC pin 8 on 16F84)
```

REVERSE *pin*

This command reverses the direction of the port or pin in the TRIS register. If a port was an output, it is changed to an input. If it was an input, then it's changed to an output.

Example:

```
REVERSE 2   'Change direction of PORTB bit 2. (PIC pin 8 on 16F84)
```

POT *pin, scale, var*

The POT command was developed to allow analog-to-digital (A/D) measurement with a standard PIC I/O pin. Some PICs have built-in A/D ports, which in my opinion is the best way to measure an analog signal. Although an A/D port is far more accurate, you may want to use the POT command at some point so I'll explain how this command works.

In resistor and capacitor circuits, the rate of charge to reach a known voltage level in the cap is based on the values of the resistor and capacitor. If you instead know the charge time and the capacitor value, then you can figure out the resistance. That's how the POT command works.

It uses the I/O pins' high and low thresholds as the trigger points for measuring the capacitor charging. The capacitor and resistor are connected to an I/O pin as seen in Figure 2-1.

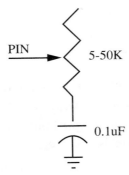

Figure 2-1: Circuit configuration for measuring capacitor charging.

When the command is processed, the capacitor is first discharged by the I/O port, which is configured by the POT command as an output and low. After that, the I/O port is changed to an input and starts timing how long it takes for the capacitor to charge up to the high threshold voltage threshold of the PIC I/O port. When that high threshold is met, the charge time is known. That charge time is converted into a 0–255 decimal value based on the value of the scale variable, where 255 is the maximum resistance and 0 is minimum.

The key is the proper scale value. It must be specified for this command to work properly. In order to have the scale value match the resistance range you are using, it must first be calculated for the R/C attached. No math is required because it must be determined experimentally. First set the resistance to its maximum value. Then set scale to 255 and run the command. The variable value returned will be the proper scale value for that R/C combination.

Example:

```
POT    3, 240, B0    ' Measure the resistance and place the 0-255
                     ' value in B0
                     ' The 240 value was found first by setting scale
                     ' to 255
```

BUTTON *pin, down, delay, rate, var, action, label*

This command is designed to make it easier to check the status of a switch. I find it very confusing, and I'm not alone! Let's examine it.

This command actually operates in a loop. It continually samples the pin and filters it for debounce. It also compares the number of loops completed with the switch closed to see if auto-repeat of the command action should take place. The auto-repeat is just like the keyboard on a personal computer. Hold down a key down, and it will soon auto-repeat that character on the screen until it runs out of space.

The command has several operators that affect its operation.

pin

This is the I/O port pin the switch is connected to as seen in Figure 2-2.

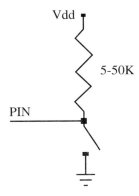

Figure 2-2: I/O port pin connection.

down

This defines what the port should see when the switch is closed, a high (1) or low (0).

delay

This is a value of 0-255 that tells the command how many loops must occur with the key pressed before starting the auto-repeat feature. This operator also does two other functions. If the value is 0, then debounce and auto-repeat are shut off. If it's 255, then debounce is on but auto-repeat is off.

rate

This value sets how fast the auto-repeat actually repeats itself. In other words, it's the rate of auto-repeat. It requires a 0–255 value.

var

This must be a variable like B0 because it stores the number of loops completed in the BUTTON command. It must be reset to zero prior to running this command or the BUTTON command will not function properly.

action

This tells the BUTTON command which state the switch must be in to jump to the location described by *label*. If you want to jump to the *label* routine when the switch is closed (as defined by *down*), then set action to 1. If you want to jump when the switch is open, then set action to 0.

label

This sets the goto label if the action operator is met. This label must be defined somewhere in the program to properly compile.

Example:

```
B0 = 0
BUTTON 2, 0, 100, 10, B0, 0, SKIP    ' Check for button press (0 at
                                     ' I/O port)at port pin 2 and
                                     ' goto SKIP routine if not
                                     ' pressed. Also if it's pressed
                                     ' and held for 100 loops,
                                     ' auto-repeat at a rate of 10
```

What makes this command so confusing is all the options. I would have preferred a simple BUTTON command with just action and label with modifiable switch debounce. Auto-repeat could have been a command on its own. I'll show examples later of how to read switches with other techniques.

This completes the I/O control section of the PBC language. Now let's look at some more familiar BASIC commands in the section I call "redirection."

Redirection

This group contains the commands used to jump around within your PBC program. This can be confusing to the beginner but anyone who has programmed before knows the power of redirection. It allows multiple options within a program all based on the logic within the PBC program structure.

GOTO *label*

This is the simplest of the bunch. It simply redirects the current program location to a new location. This can be used for bypassing a section of code accessed by another part of the program or even jumping back to the start of the program. The *label* must be defined somewhere else in the program.

Example:

```
GOTO  START ' Jump to the beginning of the program at label START
```

IF *comp* {AND/OR *comp*} THEN *label*

This command could be considered a conditional GOTO. If you have written any BASIC code then you're probably familiar with this command. The bracketed AND/OR is an optional part of the command. The *comp* term(s) is the expression that is tested. The expression must contain a variable that is compared to a constant or another variable. The expressions may use any combination of the following:

<	less than
>	greater than
=	equal to
<>	not equal to
<=	less than or equal to
>=	greater than or equal to

All comparisons are unsigned, meaning PBC can't tell the difference between a negative number or a positive number. They are all treated as absolute values. When the *comp* expression is true, the command jumps to the *label* following THEN. If the *comp* expression is not true, then the PBC command following the IF THEN command will be executed.

Example:

```
IF B0 > 10 THEN BEGIN      ' If the variable B0 is greater than 10
                           ' then jump to BEGIN

IF B0 => 10 AND B0 <= 20 THEN test   ' B0 must be less or equal to
                                     ' 20 and greater than or equal
                                     ' to 10 to jump to test
```

BRANCH *offset, (label, {label, label, ...})*

This command is a multiple level IF THEN. It will jump to the program *label* based on the *offset* value. *Offset* is a program variable. If *offset* equals zero, the program will jump to the first listed *label*. If offset is one, then the program will jump to the second listed *label*.

If *offset* is a larger number than the number of *labels*, then the BRANCH instruction will not be executed and the PBC command following BRANCH will execute.

Example:

```
BRANCH  B1, (first, second, third)   ' If  B1=0 then goto first; if
                                     ' B1=1 then goto second; if
                                     ' B1=2 then goto third; if
                                     ' B1 > 2 then skip BRANCH
                                     ' instruction
```

GOSUB *label*

This command is a temporary GOTO. Just like GOTO, it jumps to the defined *label*. Unlike GOTO, it returns back and continues with the next command after GOSUB.

GOSUB is really an abbreviation for GOto SUBroutine. A subroutine is a program listing within a main program. You can have several subroutines that each perform a special function. You can also place a common routine in one subroutine rather than write the common routine multiple times. This is a way to save memory.

You can also GOSUB within a subroutine. The first return will bring you back to the subroutine and the second return will bring you back to the original GOSUB. This is known as *nesting*. You are limited to four levels of nesting with PBC or, in other words, a maximum of four GOSUB commands may be used together.

The return is performed by an accompanying command RETURN. They must both be in the program to make the function work. You can have multiple GOSUB commands jumping to the same routine but only have one RETURN command at the end of the subroutine. This is quite common.

Example:

```
FLASH:
      GOSUB   SUB        ' Jump to subroutine SUB
      GOTO    FLASH      ' Loop again to flash LED on PORTB bit 4


SUB:
      TOGGLE  4          ' Change state of PORTB bit 4
      RETURN             ' Return to command after gosub
```

RETURN

As explained above, this command is used at the end of a PBC subroutine to return to the command following the GOSUB command.

Example:

```
Subrout:
      B0 = B0 + 1
      RETURN
```

This completes the redirection section of the PBC language. Now let's look at some of the special function commands.

Special Function

This is a group of commands with a very diverse set of functions. They are really handy commands and begin to show how easy PBC makes programming.

SOUND *pin,(note, duration {, note, duration})*

This command was created to make sounds from a PIC. A PIC alone cannot produce sound so additional hardware is required, as shown in Figure 2-3.

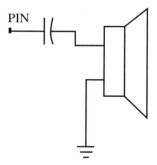

Figure 2-3: Circuit for generating sound with a PIC microcontroller.

What SOUND does is pulse the designated *pin* high and low at an audible frequency. The pulsing will continue for a length of time specified by the *duration* value. The values do not specifically tie into musical note values. The sounds produced fall into two categories, tones and white noise.

Tones are selected by the *note* value. The note value can range from 1 to 127 for tones and the higher-frequency white noise are values 128 to 255. Value 0 is for silence. It can be used to produce a pause between notes or white noise.

Duration is a value of 0 to 255 measured in milliseconds. Additional notes and duration values can be include in a single command. With the right combination, even a short melody can be produced. Using just a single note and duration makes it easy to produce feedback if a button is pressed. Here's a short program example; I'll have more examples in the later chapters.

Example:

```
SOUND 0, (100, 10, 50, 20, 100, 10, 50, 20) 'Make a cycling sound
                                             ' that alternates
                                             ' between note 100 and
                                             ' note 50 on PORTB pin
                                             ' 0. Each note has a
                                             ' different duration
```

FOR ... NEXT

This command is familiar to anyone who has used BASIC. The format is as follows:

> FOR *variable* = *start* TO *end* [STEP [-] *increment*]
>   [PBC Routine]
> NEXT {*variable*}

The PBC routine trapped between the FOR / NEXT command structure will be executed while the logical statement following the FOR command is within the *start* and *end* values.

Variable can be any variable you create with the SYMBOL command mentioned earlier. *Start* and *end* are limited to the size of the *variable*. If the *variable* is a byte then *start* and *end* must be 255 or less. If *variable* is a word size then *start* and *end* must be less than 65536.

What this command really does is first initialize the *variable* to the *start* value. It then executes the PBC Routine. At the end of the routine it increments the *variable* by one and compares it to the *end* value. If *variable* is equal to or greater than the *end* value, then the PBC command that appears after the NEXT command is executed. If *variable* is less than the *end* value then the trapped PBC routine is executed again.

The STEP option allows the command to do something other than increment the *variable* by one. It will instead increment the *variable* by the value *increment*. If *increment* is a negative number, then the *variable* is actually decremented. If a negative number is used, you must make sure *start* is a greater number than *end*.

The *variable* name after NEXT is optional. It will increment the closest FOR *variable*. If you have a FOR ... NEXT loop within a FOR ... NEXT loop, then it's best to place the proper *variable* name after the NEXT.

Here is an example of FOR ... NEXT teamed up with SOUND:

Example:

```
FOR B0 = 1 to 100              'Continue producing sound on PORT B
pin 2
        SOUND 2, ( B0, 50)     ' in 50 msec increments.
                             ' The sound will increase in pitch
NEXT                           ' with every loop until sound value
                             ' 100 is produced
```

LOOKDOWN *search,(constant {, constant}), var*

It can be difficult to remember exactly what this command does. I still look it up in the manual almost every time I use it. What it does is look down a list of values (*constant*) and compare each value to a master value (*search*). If a match is found, then the position is stored in a variable (*var*). It provides a lookup-table method for converting any character into a numeric value from 0 to 255.

If *search* matches the first *constant* then *var* is set to 0. If the second *constant* matches *search*, then *var* is set to 1, etc. String constants and numeric constants can both be part of the table.

The PBC separates the list of constants by looking at each 8-bit value. It's best to separate the constants with commas so the compiler knows where to start and where to stop. 1010 is not treated the same as 10,10. If you use string constants, then they will be treated as their respective 8-bit value. Therefore, commas may not be needed for string variables.

Example:

```
LOOKDOWN  B0,(0, 1, 2, 4, 8, 16, 32, 64, 128), B1   ' B1 contains
                                          ' in decimal which single
                                          ' bit is set in B0. If
                                          ' B0 = 128 or 10000000
                                          ' binary then B1 = 8.
                                          ' If more than one
                                          ' bit is set in B0 then
                                          ' B1 = 0
```

LOOKUP *index,(constant {, constant}), variable*

This command performs a lookup table function. *Index* is an 8-bit variable that is used to choose a value from the list of *constants*. The selected *constant* is then stored in the *variable* following the list of constants.

If the *index* variable is 0, the first constant is stored in the *variable*. If *index* is 1, then the second *constant* is stored in *variable*, and so on. If *index* is a value larger than the number of listed *constants*, then *variable* is left unchanged. The *constants* can be numeric or string constants. Each *constant* should be separated by a comma.

Example:

```
FOR B0 = 0 to 7                                    'Convert
                                                   'decimal number to
LOOKUP  B0,(0, 1, 2, 4, 8, 16, 32, 64, 128), B1   ' a single bit to
                                                   'be set
NEXT
```

PEEK *address, var*
POKE *address, var*

These commands do not come from the original BASIC Stamp language. They are unique for the PIC only and very useful. With these commands, you can access any register in the PIC and read the value or write a value at that location. This is useful for accessing other I/O ports, such as Port A, and also for reading A/D values on PICs with A/D ports. It can also be used to set up the option or status registers if you are into advanced PIC control.

Address is the location within the PIC that you want to read from (peek) or write to (poke). The *var* is the variable that contains the data to be written when using the Poke command. The *var* is the variable where the data is stored when using the Peek command.

Here is an example accessing additional I/O in port A by using both commands.

```
symbol      PORTA = 5           'PortA data register memory location
symbol      TRISA = $85         'PortA Tris register memory location

init:
      poke TRISA, 255           'Make all ports inputs
loop:
      peek PORTA, B0            'Read the signals on PORTA, store in
                                'B0
      if B0 = 5 then end        'If PORTA = %xxx00101 binary then
                                'stop the program. (xxx are
                                'unavailable pins on port A)
      goto loop                 'test again
```

RANDOM *var*

This command produces a pseudo-random number for various applications. The *var* variable must be a word variable. It will produce a value from 1 to 65535 but will not produce zero. You cannot use a port number or port variable.

Example:

```
loop:
      random  W2              ' Create a random number
      pause 100               ' pause 100 msec
      goto loop               ' do it again
```

Pulse Control

This group of commands is used to control the digital waveforms many projects require. To create a pulse requires the PIC to simply switch the I/O port from a low state to a high state and then back to low again. These commands make it much easier to do that and also receive pulses from other sources and measure the pulse width. Even digital-to-analog conversion can be accomplished if you can control the pulse width. These commands are very useful.

PULSIN *pin, state, var*

This command is great for measuring the pulse width of any signal coming into a PIC port. With the 4-MHz crystal or resonator, PULSIN will measure in 10 microsecond resolution.

The variable *pin* is a value of 0 to 7 representing the PORTB pin you want to monitor. The *state* variable determines if the high portion or the low portion of the signal should be measured. If *state* is 0 the low portion is measured. If *state* is 1 the high portion is measured.

The *var* variable is where the results are stored. If you want to measure from 0 to 2550 microsecond, then *var* could be a byte variable like B0. If you want to measure up to 655,350 microsecond, then use a word variable or W1.

Example:

```
meas:
      pulsin 3,1,w3           '  measure the high time of signal
                              ' on portB pin 3
      if w3 > 100 then warn   'test high time value if its greater
                              ' than 1 msec
      low   0                 'clear pin 0 to turn off LED
      goto meas
warn:
```

```
    high  0              ' set pin 0 high to light LED (greater than 1
                         ' msec warning)
    goto meas
```

PULSOUT *pin, period*

This command generates a single pulse from any of the PORTB pins. The *pin* variable is the PORTB pin to use. The *period* variable is the length value of the generated pulse (1 to 65535). The resolution is in 10 microsecond units so the maximum pulse width is 655,350 microseconds wide.

The pulse is generated by toggling the pin twice. Thus, the initial state of the pin determines if the pulse is high or low. It's best to set the pin to the desired state before issuing this command.

Example:

```
pulse:
    low 1                ' initialize pin1 to zero
    pulsout1, 300        ' send a high pulse 3 msec wide out
                         ' portB pin 1
    pause 10             ' pause 10 msec and do it again
    goto pulse
```

PWM *pin, duty, cycle*

This command can be used for various tasks, but a common task is creating an analog output from a digital signal. This command works slightly different than you might initially think. The command sends a series of pulses from the specified pin for a specific period of time. The pulse width of each pulse is actually fixed but the number of times the pulse is sent controls the high time versus the low time. This is how the pulse width modulation is controlled.

The *pin* variable sets which PORTB pin to send from. The *duty* variable sets the duty cycle or actually the number of times the single pulse is repeated. It can vary from 0 (0%) to 255 (100%). The *cycle* variable sets how many times the series of pulses are repeated or number of cycles.

To use this command as a digital-to-analog converter you have to connect the output to a resistor and the resistor to a capacitor. The capacitor is connected to ground. The voltage across the capacitor will vary by how many pulses or (duty cycle) that PWM produces. The example below and Figure 2-4 demonstrate this.

```
loop:
        for B0 = 0 to 255        ' Change duty cycle from 0 to 100%.
        pwm    7, B0, 150        ' Send varying duty cycle for 150
                                 ' cycles long Analog out voltage
                                 ' will slowly increase.
        next                     ' Next duty cycle.
        goto loop                ' Repeat.
```

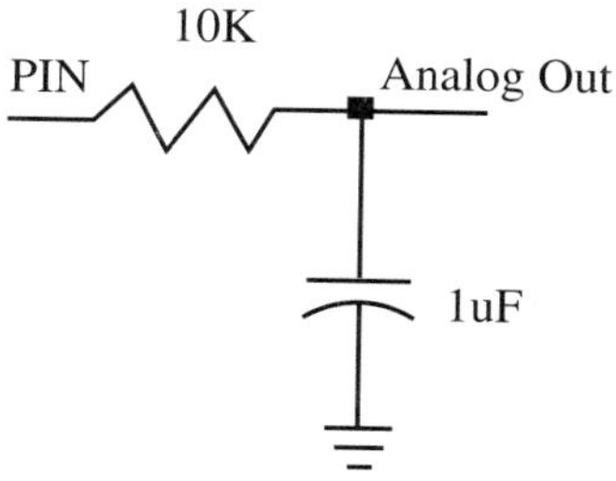

Figure 2-4: Circuit configuration to use the PWM command for analog to digital conversion.

Communication

This category of commands is exciting for the computer novice. With these single line commands, you can create PBC programs that allow a PIC to communicate with another PIC or even a PC. Anything that is RS232 compatible will most likely be capable of communicating with a PIC by using these commands. There are also commands for communicating in other signal formats.

SERIN *pin, mode, (qual, qual), (#) item, item, ...*

This command emulates the RS232 communication common on PCs, also known as serial communication. With this command many interesting programs are possible.

The command receives data from the sending source in 8N1 format, which means eight data bits, no parity, and one stop bit. The *pin* variable is the PORTB pin used. The *mode* variable is the baud rate to communicate at per the chart below.

This chart is slightly different from the BASIC Stamp because it allows 9600-baud communication in place of the Stamp's 600 baud. This is possible because a PIC programmed with PBC will run 15 times faster than a BASIC Stamp.

Here are the *mode* options:

Mode value	Baud Rate	Format
T2400 or 0	2400	TTL True
T1200 or 1	1200	TTL True
T9600 or 2	9600	TTL True
T300 or 3	300	TTL True
N2400 or 4	2400	TTL Inverted
N1200 or 5	1200	TTL Inverted
N9600 or 6	9600	TTL Inverted
N300 or 7	300	TTL Inverted

The *item* variable is the byte value received in the 8N1 format. If more than one *item* variable is listed in the command then the program will wait for the exact number of *items* listed to be received. This can lock up a program while it waits for variables. Care must be taken when using this command so you don't lock up.

The *qual* option is not needed but, if used, sets a prerequisite before accepting any *items*. The *qual* value can be a constant, variable or a string constant. The command looks for the *qual* to be received before going further.

The *item* variable can be preceded with a # character. This will convert any decimal number received into the ASCII equivalent and store that in the *item* variable. Any non-decimal values received will be ignored when using the #.

Example:

```
loop:
      serin 1, n9600, ("A"), B0          ' Wait until the ASCII value
                                         ' for A is received on portB
                                         ' pin 1 and then store the next
                                         ' byte in B0
      goto loop
```

SEROUT *pin, mode, (item, item, ...)*

This commands sends a byte or bytes in serial 8N1 format out a specified pin. The *pin* variable defines the PORTB pin used for communication. The *mode* value determines the communication baud rate. The chart below defines the mode options.

Mode value	Baud Rate	Format
T2400 or 0	2400	TTL True
T1200 or 1	1200	TTL True
T9600 or 2	9600	TTL True
T300 or 3	300	TTL True
N2400 or 4	2400	TTL Inverted
N1200 or 5	1200	TTL Inverted
N9600 or 6	9600	TTL Inverted
N300 or 7	300	TTL Inverted
OT2400 or 8	2400	Open Drain
OT1200 or 9	1200	Open Drain
OT9600 or 10	9600	Open Drain
OT300 or 11	300	Open Drain
N2400 or 12	2400	Open Collector
N1200 or 13	1200	Open Collector
N9600 or 14	9600	Open Collector
N300 or 15	300	Open Collector

The item *value*(s) can be in three formats and they can be mixed.

1) A string constant is sent as a string of characters, i.e. "hello" is sent as five individual bytes.

2) A numeric value can be sent as the ASCII equivalent (i.e., 13 will represent the ASCII carriage return and 10 will be received as a line feed). If you send the numeric value to another PIC, though, it will be received as the binary value.

3) A numeric value preceded by a # symbol will break up the number and send it as individual ASCII characters, i.e., #123 will be sent as "1", "2", and "3".

Example:

```
loop:
        for b1 = 0 to 9              ' Send 10 numbers
        serout 5, n2400, (#b1, 10)   ' 2400 baud inverted, send
                                     ' ASCII value of b1
        next                         ' followed by a line feed.
        goto loop
```

I2CIN *control, address, var {, var}*
I2COUT *control, address, var {, var}*

These commands are used to communicate with other components in the Phillips I²C format. I2CIN receives byte values and stores them in the *var*(s) variables and I2COUT sends the *var* data. They are very useful for communicating with other components such as serial EEPROM.

The lower seven bits of the *control* variable contain the control code and the chip select or additional information. This depends on the device. The high-order bit in *control* is used as a flag to indicate whether the *address* is to be sent as a 16-bit value or an 8-bit value. If that bit is 1, then it's sent as 16 bits; if 0, it's sent as eight bits.

The *address* is the location to read from or write to. For example, when communicating with a single 24LC01B 128 byte serial EEPROM the address needs to be right bits, and the chip select is unused. The *control* byte would be %01010000 for that part. (See the *Microchip Non-Volatile Memory Products Data Book* for more info on the serial EEPROM memory chips.)

These commands are also unique in that they use PORTA pins 0 and 1 for data and clock, respectively, instead of the usual PORTB pins.

Example:

```
' I2CIN and I2COUT Commands
'
' Write address to the first 16 locations of an external serial
' EEPROM
' Read first 16 locations back and send to serial out repeatedly

SymbolSO = 0                            ' Serial Output

        For B0 = 0 To 15                ' Loop 16 times
            I2Cout $50,B0,(B0)          ' Write each location's
                                        ' address to itself
            Pause 10                    ' Delay 10ms after each
                                        ' write
        Next B0

Loop:   For B0 = 0 To 15 step 2                 ' Loop 8 times
            I2Cin $50,B0,B1,B2                  ' Read 2
                                               ' locations in
                                               ' a row
            Serout SO,N2400,(#B1," ",#B2," ")  ' Print 2
                                               ' locations
        Next B0

        Serout SO,N2400,(10)            ' Print linefeed

        Goto Loop
```

Timing

This popular group of commands involves time. They really don't do anything special except waste a specified amount of time and then let the program continue or stop the program altogether. The accuracy is not something to set your watch by, but for most applications they are accurate enough.

Some of the commands put the PIC in low power mode while they waste time. This is great for battery applications.

NAP *period*

This command places the PIC in a low power mode for short periods of time.

This can be used to save power in battery applications; nap for a short time, then go check the I/O, then nap again, etc..

The *period* variable is a number from 0 to 7. Each corresponds to the delay listed below. The timing is derived from the Watchdog Timer (WDT) inside the PIC. You must have the WDT turned on when you program the PIC to use this. The WDT is driven by an internal RC circuit so its accuracy is not great. All time values are nominal but can vary as much as 20% over a range of operating temperatures.

period	delay (approx)
0	18 millisecond (msec)
1	36 msec
2	72 msec
3	144 msec
4	288 msec
5	576 msec
6	1.152 sec
7	2.304 msec

Example:

```
loop:
        if in0 = 0 then prog     ' Test pin 0 if its low
        nap 6                        ' low power for 1.152 sec delay
        goto loop                    ' test pin 0 again

prog:
        toggle 1                     ' pin 0 low,  toggle pin 1
        goto loop                    ' test pin 0 again
```

PAUSE *period*

This is one of the very useful commands. It can pause program execution for a *period* of 1 to 65,535 milliseconds. It doesn't put the PIC in low power mode, but is more accurate than the NAP or SLEEP command. You can use it to control timing or pulse widths or whatever your program requires.

Example:

```
pulse:
        high 0          'send high signal out pin 0
        pause 10        'pulse width is 10 msec plus time to execute
                        'pause command
        low 0           'send low pulse out pin 0
        pause 10        'pulse width is 10 msec plus time to execute
                        'pause command
        goto pulse      'loop again to make square wave of close to 50%
                        'duty cycle
```

SLEEP *period*

This command places the PIC in low-current mode and stops the PIC from running for a length of time. The *period* variable sets the amount of time to stay in the low-power mode.

The SLEEP command timing is controlled by the watchdog timer (WDT) within the PIC, which is just an RC circuit. To make the SLEEP timing more accurate it is compared to the system clock every 10 minutes and adjusted. The *period* value can range from 1 to 65535 and represents increments of 2.3 seconds. A value of 1 will make the PIC sleep for 2.3 seconds, while a value of 65535 will make the PIC sleep for just over 18 hours.

Example:

```
loop:
        if in0 = 0 then prog    ' Test pin 0 if its low
        sleep 26                ' low power for 1 minute delay
        goto loop               ' test pin 0 again

prog:
        toggle 1                ' pin 0 low,  toggle pin 1
        goto loop               ' test pin 0 again
```

END

This command stops program execution. It is automatically placed at the end of your assembled PBC program if you don't include it, but it's best to include it at the end of your main loop. Place the subroutines after this command. You should never actually get to this command since your PIC just stops and goes into an endless series of NAP commands and never leaves until you reset the PIC.

Memory

The following commands only work on PICs with internal EEPROM memory. At the time this book was written, that included the 16C8X, 16F8X, and 16F87X devices. Microchip is expanding the EEPROM family so more choices will soon be available.

These memory commands are great for storing measured data or constants you don't want to lose when the power goes out. That's because EEPROM memory is nonvolatile. Data stored in EEPROM can last as long as ten years without power. Let's start with the READ command.

READ *address, var*

This command will only work on PICs with internal EEPROM like the 16F84. The *address* variable is the location to read. The value read at the specified location will be stored in the *var* variable. The command also has a special mode. If *address* is the value 255 then the total number of space available in the PIC EEPROM will be put in *var*.

Example:

```
loop:
        read 10, b1        'Store value at location 10 in variable b1
        goto loop          ' repeat
```

WRITE *address, value*

This command will only work on PICs with internal EEPROM like the 16F84. The *address* variable is the location to write to. The value in the *var* variable will be stored at the specified location.

Example:

```
loop:
        for b1 = 0 to 9        ' The first 10 locations in EEPROM
        write b1, 0            ' are set to 0.
        next
```

EEPROM *location, (constant, constant)*

This command is different than any of the other commands because it is only executed when the PIC is programmed, not when the program in the PIC is run. It is used to preload the EEPROM memory.

The *location* variable *is* optional and sets the starting point to store values. If it is not included in the command then the location 0 is used. The *constant* value can be a numeric or string constant. Only the least significant byte is stored for numeric constants and string constants are treated as separate ASCII values.

Example:

```
EEPROM 5, (10,"A")          'Store value 10 at location 5, ASCII value
                            'of "A" at location 6 of the internal
                            'EEPROM
```

Assembly Language

At some point you will want to do something that PBC can't do, or at least something that would be easier or quicker using assembly language. Assembly language is a little more difficult to understand, and most of your programs will consist of only PBC. But it's nice to have the option of sticking a few lines of assembly in when you need it. These are the commands that allow that.

I also suggest you read the back sections of the PBC manual before using these commands. The manual explains how the compiler handles assembly and gives tips on the best way to mix assembly and PBC.

ASM ... ENDASM

These commands are used together to insert assembly language into a PBC program. This is handy for simple things or more control over how long a command executes. You should read the PIC manual before dealing with assembly language. Comments must start with a semicolon when commenting assembly commands.

Example:

```
asm                    'The following code is written in assembly
_assembly              ;Label must be proceeded by underscore
      clrb   RP0       ; Comments must be proceeded by semicolon
      mov    5,_B0     ; for assembly instead of the single quote for
                       ; PicBasic
endasm
```

CALL *label*

Similar to GOSUB, except this executes an assembly language subroutine rather than a PBC subroutine. It can be tricky to use this and it should only be used when your program just HAS to use assembly.

Example:

```
CALL   assembly

asm                    'The following code is written in assembly
_assembly              ;Label must be proceeded by underscore
      clrb   RP0       ; Comments must be proceeded by semicolon
      mov    5,_B0     ; for assembly instead of the single quote for
                       ; PicBasic
endasm
```

Command Summary

Programming in PBC is quite easy and is only limited by your imagination and the amount of PIC memory you have. I only included short snippets of code examples in this chapter, but later in this book I'll cover actual applications.

When you purchase the PBC complier, you also get a manual that covers these commands in similar detail. Hopefully, I've filled in any gaps the manual has and at least given you a thorough introduction to the PBC language. Now let me explain how to actually use this compiler.

Using PBC

The PBC is really a DOS program that can be run from a true DOS prompt or within a DOS window on a Windows machine. The command format is very simple. The tough part is writing your program.

You create your program with any text editor you like and when it's completed, you name the file any name with eight or fewer characters with a ".bas" suffix. The PBC defaults to looking for ".bas" files. After you have created the file, you then invoke the PBC with the following command:

PBC *filename*.bas

filename is the name you gave your program.

The PBC then goes into action. It first compiles your file into an assembly file. The output file should end up in the same directory as the original *filename*.bas. The assembly file will be called *filename*.asm.

That ".asm" file is then assembled by the integral assembler into a file named *filename*.hex. That ".hex" file is the 1's and 0's version of your program that actually is used to program the PIC.

This whole process takes just a few minutes, unless you have errors. If errors are present in your PBC program, the compiler briefly describes the error and what line it's on. You then have to go back and fix what is wrong and compile again.

It's very rare to write a program the first time without errors, so don't get discouraged when you have errors. Hopefully, the program examples in the later chapters will help guide you through successful code development.

Recently, several users of PBC and PBPro have developed Windows editors designed for using PBC or PBPro. These make the above steps even easier since it's all done within the same screen.

The PBC also has several command line options. I skipped that previously for the sake of simplicity, but let's talk about that now.

Options

The complete PBC command line is as follows:

PBC *-options filename*.bas

The *-option* is the list of command line options you may want to use to control how the PBC works on your file. The options are just that, optional. All options must be preceded by a "-" symbol. Here are the options:

option -C

This limits the compile to just the BASIC Stamp (BS) command set. It doesn't allow any more variables than the BS allows and it doesn't allow any of the PBC commands that the BS doesn't have. This option makes PBC completely compatible, forward and backward, with the BS.

option -D

This causes the assembler to generate a symbols table, a listing file, and a map file in addition to the .hex file.

option -P

This option is for programming PICs other than the default 16F84.

option -L

This creates a listing file in addition to the .hex file.

option -OB

This forces PBC assembler to produce a binary file instead of the Merged Intel HEX file. It's mainly for older programmers.

option -Q

This stops the default .bas and forces you to include the full filename suffix so you can use something other than .bas if you desire.

option -S

This prevents the assembler from invoking leaving just the assembly file and not a .hex file. You then have to assemble the resultant file.

As you can see, the PBC is a full-feature compiler despite its easy-to-use BASIC language core. PBC is a great way to start and most hobbyists will find it's all they need. To fully utilize PBC, it helps to understand the inner workings of the PIC processors. Because the PICs use a common architecture, once you learn one PIC your knowledge carries over to any other PIC you may want to use. Once you learn the inner workings you may want to upgrade to the PBCPro compiler or PBPro. Let's look at that before we start using these compilers.

CHAPTER 3

The PicBasic Pro Compiler

If you're really serious about PICs and want to stay with PicBasic, microEngineering Labs has developed a professional version of the PicBasic compiler called PicBasic Pro (PBPro). microEngineering Labs had so many things they wanted to add to PBC that it became a BASIC compiler in a league by itself. PBPro added many more features to PBC. The main benefits of PBPro over the PBC are:

- Interrupts in BASIC
- Programs longer than 2k
- Arrays
- Direct access to all I/O without using PEEK and POKE
- Direct access to special function registers without using PEEK and POKE
- Ability to tell Pro what clock oscillator you want to operate at instead of the 4 MHz PBC expects
- Ability to use PBPro with Microchip's assembler MPASM for better ICE and simulator compatibility
- Any variable can be designated as bit, byte, or word.

PBPro is really a full-featured compiler. By the time you read this, microEngineering Labs should have announced the release of additional support for the 17CXXX high-performance 16-bit core PICs. Here is a list of the current PBPro commands:

@: Insert one line of assembly language code.

ASM..ENDASM: Insert assembly language code section.

ADCIN: Read on-chip analog to digital converter.

BRANCH: Computed GOTO (equivalent to ON..GOTO).

BRANCHL: BRANCH out of page (long BRANCH).

BUTTON: Debounce and auto-repeat input on specified pin.

CALL: Call assembly language subroutine.

CLEAR: Zero all variables.

CLEARWDT: Clear Watchdog Timer.

COUNT: Count number of pulses on a pin.

DATA: Define initial contents of on-chip EEPROM.

DEBUG: Asynchronous serial output to fixed pin and baud.

DEBUGIN: Asynchronous serial input from fixed pin and baud.

DISABLE: Disable ON INTERRUPT processing.

DISABLE DEBUG: Disable ON DEBUG processing.

DISABLE INTERRUPT: Disable ON INTERRUPT processing.

DTMFOUT: Produce touch-tones on a pin.

EEPROM: Define initial contents of on-chip EEPROM.

ENABLE: Enable ON INTERRUPT processing.

ENABLE DEBUG: Enable ON DEBUG processing.

ENABLE INTERRUPT: Enable ON INTERRUPT processing.

END: Stop execution and enter low power mode.

FOR..NEXT: Repeatedly execute statements.

FREQOUT: Produce up to 2 frequencies on a pin.

GOSUB: Call BASIC subroutine at specified label.

GOTO: Continue execution at specified label.

HIGH: Make pin output high.

HSERIN: Hardware asynchronous serial input.

HSEROUT: Hardware asynchronous serial output.

I2CREAD: Read bytes from I^{2}C device.

I2CWRITE: Write bytes to I^{2}C device.

IF..THEN..ELSE..ENDIF: Conditionally execute statements.

INPUT: Make pin an input.

{LET}: Assign result of an expression to a variable.

LCDIN: Read RAM on LCD.

LCDOUT: Display characters on LCD.

LOOKDOWN: Search constant table for value.

LOOKDOWN2: Search constant/variable table for value.

LOOKUP: Fetch constant value from table.

LOOKUP2: Fetch constant/variable value from table.

LOW: Make pin output low.

NAP: Power down processor for short period of time.

ON INTERRUPT: Execute BASIC subroutine on an interrupt.

OUTPUT: Make pin an output.

PAUSE: Delay (1 millisecond (msec) resolution).

PAUSEUS: Delay (1 microsecond (_sec) resolution).

PEEK: Read byte from register.

POKE: Write byte to register.

POT: Read potentiometer on specified pin.

PULSIN: Measure pulse width on a pin.

PULSOUT: Generate pulse to a pin.

PWM: Output pulse width modulated pulse train to pin.

RANDOM: Generate pseudo-random number.

RCTIME: Measure pulse width on a pin.

READ: Read byte from on-chip EEPROM.

READCODE: Read word from code memory.

RESUME: Continue execution after interrupt handling.

RETURN: Continue execution at statement following last executed GOSUB.

REVERSE: Make output pin an input or an input pin an output.

SERIN: Asynchronous serial input (8N1) (BS1 style with timeout.)

SERIN2: Asynchronous serial input (BS2 style.)

SEROUT: Asynchronous serial output (8N1) (BS1 style.)

SEROUT2: Asynchronous serial output (BS2 style.)

SHIFTIN: Synchronous serial input.

SHIFTOUT: Synchronous serial output.

SLEEP: Power down processor for a period of time.

SOUND: Generate tone or white noise on specified pin.

STOP: Stop program execution.

SWAP: Exchange the values of two variables.

TOGGLE: Make pin output and toggle state.

WHILE..WEND: Execute code while condition is true.

WRITE: Write byte to on-chip EEPROM.

WRITECODE: Write word to code memory.

XIN: X-10 input.

XOUT: X-10 output.

As you can see, the PBPro list of commands is extensive. Some of the commands you will use often and other commands will be specific to unique applications. You will find that some PBPro commands operate slightly different from PBC commands of the same name and some commands function the same but have a slightly different format. Commands are not the only thing different about PBPro. The way the PBPro compiler handles variables, constants, symbols, and pin names make it a more powerful compiler. Let's look at those first.

Variables

PBPro uses the RAM of the PIC as open space for storing data the same way an assembly language program would. The RAM space is not predefined with the B0 or W0 names that PBC used. Instead, you can name the RAM anything you want using the VAR directive. You also have to tell the compiler how much RAM space the variable will use. It can be defined as a bit, byte or word variable. The format is as follows:

label   VAR   *size*{.modifiers}

The *label* is the name of the variable you will use throughout your program. PBC and PBPro are case insensitive, which means LABEL and label are treated the same by both compilers. A label can have up to 32 characters, but you'll probably only use five or six. Why create such a long name if you're going to repeat it 20 or 30 times throughout a program? Save yourself the typing and make them short but understandable.

VAR is the directive that lets PBPro know that this line is establishing a variable. It doesn't have to be capitalized.

Size is the amount of RAM space the variable is going to use. It has to be bit, byte or word. Nibble is not an option.

Examples:

```
Book        var     word
Page        var     byte
Letter      var     bit
```

The *size* parameter can also be modified or specified to be a part of a bigger memory size variable. Also, the bit number and the bit word ("0" and bit0) are treated the same. Modifier names are listed below.

0 and bit0 are the 1st bit of the byte or word;

1 and bit1 are the 2nd bit if the byte or word;

and so on until. . . .

15 and bit15 are the highest or 16th bit of the byte or word.

byte0 and lowbyte are the least significant byte of the word; byte1 and highbyte are the most significant byte of the word.

Examples:

```
letter0     var     page.0          'letter0 is bit0 of the byte page
letter1     var     page.bit1       'letter1 is bit1 of the byte page
chap0  var     book.byte0      ' chap0 is the first byte of the word
                               ' variable book
chap1  var     book.highbyte   ' chap1 is the second byte of the
                               ' word variable book
```

The VAR directive can also be used to add another name or alias to the same variable.

Example:

```
Novel   var     book           'novel or book will refer to the same RAM
                               'location
```

You can also create arrays with PBPro, which is very handy for some programming routines. You use the same VAR directive and then specify the number of array elements in the size modifier. The number of elements must be with brackets.

Examples:

```
chapter          var     byte[10]
```

These will be treated as chapter[0], chapter[1], ..., chapter[10]. You can even use them in a loop.

Example:

```
for x = 1 to 10
chapter[x] = x
next
```

Because of memory limitations, arrays are limited as follows:

✎ Bit: 128 elements

✎ Byte: 64 elements

✎ Word: 32 elements

PBPro uses 24 bytes of RAM for its own internal use and can occasionally grab a few more bytes for sorting out complex equations. How many variables you can have depends on the PIC part you are using. A 16F84 PIC, which has 68 bytes of RAM, will yield (68 - 24), or about 44 byte variables.

If you still like using the B0 and W0 predefined variables, PBPro has two `include` files that you enter at the top of your program.

```
Include        "bs1defs.bas"
```

or

```
Include          "bs2defs.bas"
```

These contain numerous `var` statements that define the B0 and W0 type variables used in the Basic Stamp modules. I don't recommend them, though. You will have more space and much more control if you create variables yourself. It's also better from a programming technique because a label like "number" is far more understandable than "B0".

Constants

Constants are handled differently in PBPro. Instead of using the SYMBOL command as in PBC, you instead use the CON directive. Constants are handy for having a name reference a value so that reference can be used throughout the program rather than the actual number. This offers the opportunity to change the value once at the CON line of the program and the compiler will then change the value everywhere the constant name is used.

You can also define a constant by an equation that uses another constant, so a single value change can have a domino effect throughout your program.

Example:

```
Value   con     100
Tenth   con     value/10
Price   con     value * 2
```

As you can see, any change to the *value* constant will also change the *tenth* and *price* constants throughout the program.

Symbols

The SYMBOL command in PBC is used in PBPro but is limited to renaming variables and constants.

Examples:

```
Symbol          cost = price
Symbol          value = 100
```

Numeric and ASCII

PBPro has three ways to define a number: decimal, binary, and hexadecimal. You will find advantages to each as you program. Most BASIC language programmers like the simplicity of just using decimal numbers and PBPro defaults to that format. Any number is assumed to be decimal.

If a number is preceded by a % symbol, then the number is considered binary. Binary numbers are always assumed to be least significant bit to the right and most significant bit to the left. If less than eight bits are shown, then the missing bits are assumed to be zero and to the left of the number shown.

If the number is preceded by a $ symbol, then the number is considered a hexadecimal number. Then there are also ASCII characters, which are basically every character you can create on a computer screen. These characters are typically letters, but if the letters are enclosed in double quotes then PBPro treats the letter or character as the ASCII numeric equivalent. Here are examples of each numeric format.

```
100             'Decimal value 100
%100            'Binary equivalent to 00000100 or decimal 4
%11110000       'Binary format used to show each bits value. The
                'higher four bits 'are 1, lower are 0
$100            'Hexadecimal number equivalent to 256 decimal.
$F0             'Hexadecimal number that is the same as %11110000
                'binary.
"A"             'ASCII character that PBPro uses decimal 65, the
                'ASCII value for A
```

Strings

If a string of characters are within double quotes, PBPro breaks them up into separate ASCII characters and uses their ASCII numeric value.

Example:

```
"HELP"   ' is treated as "H", "E", "L", "P" and uses the ASCII
         ' values for each character
```

This can be handy when sending ASCII characters to a PC or some other serial module. A phrase can be put in quotes and then each letter's ASCII value will get sent as a string of bytes. You'll see more of this in the command descriptions.

I/O Access

It may seem strange that I jump to a hardware-related topic from the data format discussion, but it's because PBPro has added the feature of easily defining an I/O value the same way you would define a bit or byte value. The ports within the PIC are predefined by PBPro so access to an I/O pin on Port B is as easy as this:

```
PORTB.1 = 1             'This sets the port B bit 1 to a value of 1.
```

What it doesn't do is make PORTB.1 an output. You have to do that separately. That's the difference between setting a port high or low with this method and setting a port using the HIGH or LOW command. The HIGH and LOW commands first make the pin an output, and then sets the state. This is a great example of where PBPro separates itself from PBC and the Basic Stamp limitations. It also moves you closer to using the PIC to its full potential. It really helps to understand the inner workings of the PIC, and we'll cover that in the next chapter.

The PIC has several special function registers, and PBPro has a predefined name for each one. The list of names is found in the file PIC14EXT.bas that is included with PBPro. I recommend you print it out and add it to the PBP manual. I have mine stapled to the inside cover for easy reference. Registers like the STATUS or INTCON are used a lot so those you will remember. But others like SSPBUF or SSP-CON for the serial port peripheral on some PICs are ones you'll probably look up. The nice thing about PBPro is they all can be modified just like the example of PORTB earlier. You can define the contents of the register or location with a mathematical expression. To set the PORTB to four inputs and four outputs you would set up the TRISB register as follows:

```
TRISB = %00001111          ' RB7 - RB4 outputs; RB3 - RB0 inputs
```

This is a real time saver and a great feature of PBPro. It's also a format very similar to programming in assembly. If you ever have to write in assembly, using PBPro will have prepared you quite well.

The VAR directive has other functions. You can even rename a port pin for its function using the VAR command and then set the bit value.

For example:

```
led     var     PORTB.0    'LED is connected to port B bit 0
led = 1                     'set port B bit 0 to a high or 1 value
```

As you can see, I/O control can be as easy as defining a variable. This example assumes that PORTB is already set up as an output, otherwise the led = 1 line will set the bit but the LED won't see the value.

I/O Control

PBPro expands the available I/O that can be accessed with just a numeric reference. PBPro increased the range from the 0-7 available in PBC to 0-15. This means a command such as "HIGH 10" is legal in PBPro. Those 0-15 reserved I/O are there to maintain compatibility with the BASIC Stamp modules. If you're a Stamp user and want to use all the capabilities of PBPro, it's time to break that habit! As I described earlier, any pin on the PIC can be called by its portname.number format (PORTB.0). Commands like "HIGH 10" can be replaced with "HIGH porta.1". If you are using a PIC with more than 16 I/O, then using the portname.number format will give you direct access to every I/O with every PBPro command. If you still want to use the 0-15 digits in your commands, then you need to know which ports they work with on each PIC. Because each PIC PORT set-up is defined by its package, the following shows how the 0-15 numbers are assigned.

PIC Package	PINs 0 – 7	PINs 8 - 15
8 pin	GPIO (0 - 6)	none
18 pin	PORTB (0 - 7)	PORTA (8 - 12)
28 pin (except 14000 PIC)	PORTB (0 - 7)	PORTC (8 - 15)
28 pin 14C000 only	PORTC (0 - 7)	PORTD (8 - 15)
40 pin	PORTB (0 - 7)	PORTC (8 - 15)

As you can see, the 8-pin PICs don't use the PORTB at all but instead use the GPIO name that Microchip uses on the 8-pin PICs. Why Microchip didn't use PORTB or PORTA is a mystery to me, but if you want to define a bit in GPIO, it's also defined by PBPro in the PIC14EXT.bas file.

Example:

```
Led     var    GPIO.0  'LED connected to bit 0 of the GPIO port.
```

If you're a BASIC Stamp user, then you are familiar with the PINs command. This does the same thing PORTB.0 does but in its own way. PBPro will allow you to use the BASIC Stamp I/O format if you include the BS1 or BS2 definition files.

```
Include      "bs1defs.bas"
```

or

```
Include      "bs2defs.bas"
```

bs1defs.bas defines the Pins, B0-B13, W0-W6 names, and bs2defs.bas defines the Ins, outs, B0 - B25, W0 - W12 names.

I highly recommend you break the BASIC Stamp format habit when using PBPro. PBPro is so much more powerful. Many people, including me, started out using Basic Stamps and then switched to PBC or PBPro. If you are a Stamp user you probably have many Basic Stamp programs already written. In my opinion it's well worth the time converting those programs to pure PBPro format. Then you have a large base of PBPro programs to work from as you develop other projects.

Comments

Comments within a PBPro program can be formatted in the same two ways as PBC. The comments can be preceded by a single quote (') or the REM keyword.

```
HIGH   1      ' This would be the comment
LOW    1      REM This would also be a comment
```

Math Operators

PBPro adds a bunch of new math operators to the list PBC offers. PBPro also performs math in hierarchical order. That means multiplies and divides are done before addition and subtraction. Equations within parentheses are done first and then the rest of the equation. All math is unsigned and uses 16-bit precision, the same as PBC. MIN and MAX also perform differently than PBC. When used in PBPro, they return the minimum or maximum value rather than limit the minimum or maximum value. The following table shows the math operators.

+	Addition
–	Subtraction
*	Multiplication
**	Most significant bit (MSB) of multiplication
*/	Middle 16 bits of multiplication
/	Division
//	Division remainder only

`<<`	Shift left
`>>`	Shift right
`ABS`	Absolute value
`COS`	Cosine
`DCD`	2n decode
`DIG`	Digitize
`MIN`	Limit result to minimum value defined
`MAX`	Limit result to maximum value defined
`NCD`	Encode
`REV`	Reverse bits
`SIN`	Sine
`SQR`	Square root
`&`	Bitwise AND
`|`	Bitwise OR
`^`	Bitwise XOR
`&/`	Bitwise AND NOT
`|/`	Bitwise OR NOT
`^/`	Bitwise XOR NOT

Arithmetic Operators

Multiplication

Multiplication is actually 16×16, resulting in 32-bit results.

```
W2 = W1 * 100        ' The lower 16 bits of the result are placed
                     ' in W2

W2 = W1 ** 100       ' The upper 16 bits of the result are placed
                     ' in W2

W2 = W1 */ 100       'The middle 16 bits are stored in W2
```

Division

Division is done as follows.

```
W2 = W1 / 100    ' The numerator of the result is placed in W2

W2 = W1 / / 100       ' The remainder only is placed in W2
```

ABS

An absolute value is return up to 127 for bytes and 32767 for words. If the value is greater then 127, the result returned is $256 - $ value and words is $65536 - $ value.

```
Answer = ABS  B0              ' If B0 = 100 then Answer = 100
                             ' If B0 = 200 then Answer = 56
```

COS, SIN

These generate the cosine or sine of a number, although differently than the typical cosine/sine math. It creates a value in binary radians between -127 and $+127$, and not the typical degrees. You have to do your own conversion of binary radians to degrees. The command uses a lookup table to create the answer. If you need these kinds of calculations, PBPro has them reduced to a single command.

SQR

This operator calculates the square root of a value. However, it will only return an 8-bit integer value.

MIN, MAX

This operator is used in PBC but PBPro changed the function. PBC had MIN and MAX set the limit for a variable. PBPro measures two variables against each other and returns the minimum or maximum value. If one variable is set to a constant value, then you accomplish the same thing PBC was doing. But if both variables are changing, then these make a quick IF - THEN type function.

```
W0 = W1 MAX W2        'W0 will always equal the larger of W1 and W2
```

Binary Functions

Shift <<, >>

This is a new operator PBC doesn't have. It allows you to shift a word or byte bit by bit left or right. As bits are shifted left or right, the empty bit spaces are filled with zeros. It does not shift a 1 bit from the left all the way around to the right.

```
Result = %11110000 >> 4        'Variable Result equals
                               '%00001111 when done
Result = %00001111 << 4        'Variable Result equals
                               '%11110000 when done
```

DCD

The name is confusing but it's a handy operator. This operator sets a specific bit to 1 while clearing all the other bits. It operates on word and byte variables. The range is bits 0–15.

```
W0 = DCD 0     ' W0 will have the 1st bit set %0000000000000001
               ' when completed
```

NCD

This does the opposite of DCD. Instead of setting a bit, it reads the highest bit set and returns the position as a numeric number. If no bit is set, it returns a zero. If more than one bit is set, only the highest bit position will be returned. The number returned will range from 1–16 with 0 reserved for the no bit set answer.

```
B0 = %00010001
B1 = NCD B0           'B1 will equal 5 because the 5th bit is the
                     'highest bit set
```

REV

This operative is short for "reverse." It reverses the order of a specific number of bits. The operative is followed by the number of bits to reverse, starting at the lowest-order bit. The number to be reversed can be 1–16.

```
B1 = %11110000 REV 8        ' The result is B1 = %00001111
```

```
DIG
```

This is not really a binary function, but it is bit related. DIG returns a single digit from a group of up to five digits. If a variable has the decimal value of 54321, any one of those digits can be plucked from the number and placed by itself in a variable all its own. The choices are 0–4, with 0 being the right-most digit.

```
W0 = 54321
B1 = W0 DIG 2          'B1 = 3 or %00000011
```

Digital Operators: &, |, ^, ~, &/, |/, ^/

These are the characters that make a program look very strange to a nonprogrammer. They are just a way of doing digital logic (AND, NAND, OR, NOR, etc.) on variables. They can be handy, but hard to remember if you don't use them often.

```
B1 = %11001111
B0 = %11110000 & B1    ' B0 will AND each bit and the result will
                       ' be %11000000
B2 = %11110000 | B1    ' B0 will OR each bit and the result will be
                       ' %11111111
B3 = %11110000 ^ B1    ' B0 will Exclusive OR each bit; B0 =
                       ' %00111111
```

PBPro Commands

The PBPro list of commands is extensive, as mentioned earlier. Some of them are identical to the PBC commands and some are either not available to PBC users or have the same name as PBC commands but function slightly different. I won't group these like I did in Chapter 2, as it would be too hard to find them when you reference this later. Instead I'll list them alphabetically. I will repeat the descriptions from the PBC chapter for commands that are identical to PBPro commands.

The PBPro command structure deals with the unique PIC memory much better than PBC. PICs have an internal page boundary that is limited to 2k. PBC didn't handle this well and PBPro fixes that issue. PBPro even has some special commands for dealing with tables that cross the 2k page boundary. For the most part, though, you won't have to worry about the page boundary. The PBPro compiler takes care of that for you. Here is the descriptive list of commands:

@ *insert one line of assembly language code*

Put this symbol in front of any command and the compiler will think it's an assembly language statement or command. This makes it easy to insert a single assembly language command. PBPro has so many command options that this may not be necessary. What is handy about this command, though, is using it to insert an assembly language program. If you have a program that performs a special function, and it's written in assembly to conserve space, then just insert it with this command.

Example:

```
@        Include debugger.asm
```

ADCIN *channel, var*

This command is not the best example of PicBasic efficiency but it does save several coding steps. It was developed to make reading the analog-to-digital (A/D) ports easier but still requires so much set-up that the command really only saves a few lines of code.

The command requires the A/D *channel* you want your program to read. This can be a number from 0–7. Also required is the *variable* you want the value stored in. This should be a byte variable for 8-bit A/D and a word variable for 10- or 12-bit A/D. The number of bits in your result depends on which PIC you are using and how you configure the PIC A/D port.

I wish this command did more for your A/D set-up than it does but it can be a handy command. You see, all PICs with A/D allow the A/D ports to be configured as digital ports also. Therefore, before you use this command you have to make sure the port is properly set up for A/D input.

First you have to set the port to input mode:

```
TRISA = %11111111               ' all PORTA pins are inputs
```

Second, you have to set up the ADCON1 register. This register selects which A/D ports will be used as A/D ports and which are configured as digital ports. Consult the PIC data sheets to explain the ADCON1 register options; I also suggest you read the PIC data sheet on A/D before using this command. It will make a lot more sense to you. The ADCON1 register also controls how the result is placed in the ADRESH and ADRESL registers. The result can be right or left justified for 10- and 12-bit A/D ports. The directive to set up ADCON1 for all A/D PORTA and right justify the 10-bit result is:

```
ADCON1 = %10000010      'all PORTA analog and right justify the
                        'result
```

After the TRISA and ADCON1 steps are done, then the ADCIN command can be used:

```
ADCIN 0, value     ' Read A/D channel 0 and store the result in
                   ' variable value.
```

This command also has a few Defines that should be included with every PBPro program that uses the ADCIN command.

DEFINE ADC_BITS: used to set the result to a byte or word result. It should be followed by an 8 or a 10.

DEFINE ADC_CLOCK: used to select the internal clock source that the A/D port will use in its sample and hold circuitry. The define should be followed by a value of 0 to 3.

0: External oscillator/2

1: External oscillator/8

2: External oscillator/32

3: PIC internal RC oscillator

The most common selection here is the PIC's internal RC oscillator, which would be the value 3. Here is an example of reading a 10-bit A/D on channel 0, assuming a PIC16F876 is being used:

```
Define ADC_BITS        10      ' 10 bit result
Define ADC_CLOCK   3           ' use internal RC oscillator

advalue  VAR   word            ' Word variable to store result

TRISA = %11111111              ' All PORTA set as inputs
```

```
ADCON1 = %10000010          'all PORTA analog and right justify
                            'the result

Loop:
ADCIN 0, advalue            ' Read channel 0 and store in advalue
variable
Pause 100                   ' Pause 100 msec before taking
another sample
......... .                  ' insert code here for what you do
with the advalue
Goto loop                   ' Jump back and sample again
```

ASM..ENDASM

These commands are used together to insert assembly language into a PicBasic program. This is handy for simple things or more control over how long a command executes. (One more time: you should really read the PIC data sheet before using assembly language!) Comments must start with a semicolon when commenting assembly commands.

Example:

```
Asm                         'The following code is written in assembly
_assembly                   ;Label must be proceeded by underscore
        clrb    RP0         ; Comments must be proceeded by semicolon
        mov     5,_B0       ; for assembly instead of the single quote
                            ; for PicBasic
endasm
```

BRANCH *offset, [label, {label, label, ...}]*

This command is similar to the PBC version, except it uses brackets "[ ]" instead of "()" to frame the list of labels. The list of labels can include up to 256 different labels but they must all be on the same 2k memory page.

The BRANCH command is a multiple level IF THEN. It will jump to the program *label* based on the *offset* value. *Offset* is a program variable. If *offset* equals 0, the program will jump to the first listed *label*. If offset is 1, then the program will jump to the second listed *label*, and so forth. If offset is a larger number than the number of labels, then the branch instruction will not execute and the next PicBasic command following BRANCH will execute.

Example:

```
BRANCH  B1, [first, second, third]   ' If  B1=0 then goto first; if
                                     ' B1=1 then goto second;
                                          ' if B1=2 then goto
                                          ' third; if B1 > 2 then
                                          ' skip BRANCH
                                          ' instruction
```

BRANCHL *offset, [label, {label, label, ...}]*

This works the same as the BRANCH command except it takes care of the 2k page boundary. It also limits the number of labels to 128.

BUTTON *pin, down, delay, rate, var, action, label*

This command is designed to make reading a switch easier. I find it very confusing and I'm not alone. This command actually operates in a loop. It continually samples the pin and filters it for debounce. It also compares the number of loops completed with the switch closed to see if auto-repeat of the command action should take place. The auto-repeat is just like the keyboard on a personal computer. Hold down a key down and it will soon auto-repeat that character on the screen until it runs out of space. The command has several operators that affect its operation.

pin

This is the I/O port pin the switch is connected to as seen in Figure 3-1.

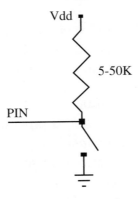

Figure 3-1: I/O port pin switch connection.

down

This defines what the port should see when the switch is closed, a high (1) or low (0).

delay

This is a value of 0–255 that tells the command how many loops must occur with the key pressed before starting the auto-repeat feature. This operator also does two other functions. If the value is 0, then debounce and auto-repeat are shut off. If it's 255, then debounce is on but auto-repeat is off.

rate

This value sets how fast the auto-repeat actually repeats itself. In other words, it's the rate of auto-repeat. It requires a 0–255 value.

var

This must be a variable because it stores the number of loops completed in the BUTTON command. It must be reset to zero prior to running this command or the BUTTON command will not function properly.

action

This tells the BUTTON command which state the switch must be in to jump to the location described by *label*. If you want to jump to the *label* routine when the switch is closed (as defined by *down*) then set action to 1. If you want to jump when the switch is open then set action to 0.

label

This sets the goto label if the action operator is met. This label must be defined somewhere in the program to properly compile.

Example:

```
B0 = 0
BUTTON2, 0, 100, 10, B0, 0, SKIP    ' Check for button press (0 at
I/O port)
                                    ' at port pin 2 and goto SKIP
                                    ' routine if not pressed. Also
                                    ' if it's pressed and held
                                    ' for 100 loops, auto-repeat
                                    ' at a rate of 10
```

What makes this command so confusing is all the options. I would have preferred a simple BUTTON command with just action and label with modifiable switch debounce. Auto-repeat could have been a command on its own. I'll show examples later of how to read switches with other techniques.

CALL *label*

This is similar to GOSUB, except this executes an assembly language subroutine rather than a PicBasic subroutine. It can be tricky to use this and should only be used when your program absolutely must use assembly.

Example:

```
CALL    assembly

Asm                         'The following code is written in assembly
_assembly                   ;Label must be proceeded by underscore
        clrb    RP0         ; Comments must be proceeded by semicolon
        mov     5,_B0       ; for assembly instead of the single quote
                            ; for PicBasic
endasm
```

CLEAR

This new command clears all RAM variables to zero. The PIC does not automatically do this, and it's a good practice to preset or clear all variables at the beginning of the program. This command makes clearing the variables easy.

Example:

```
CLEAR          'Set all variables to 0

Start:   Put your main program loop here after the CLEAR command
```

CLEARWDT

This is a command that was created in response to user feedback. If you understand the inner workings of the PIC, you know the Watchdog Timer can time out and reset your program back to the beginning. The SLEEP and NAP instructions use the Watchdog Timer to wake up the PIC after the time specified by these commands has expired. Other than that, the Watchdog is just there to reset the PIC if for some reason the program locks up. It's like an automatic "Control-Alt-Delete" function like that on a PC. The internal code of all the PBC and PBPro commands reset the Watchdog so resets don't occur when your program is running properly. This can also be done with an assembly language insert:

```
@ CLRWDT
```

Some users wanted this direct control of the watchdog reset in PICBASIC, so the CLEARWDT was developed.

Example:

```
loop:
High portb.1          ' set portb pin 1 high
Asm                   'The following code is written in assembly
_assembly             ;Label must be proceeded by underscore
       clrb    RP0    ; Comments must be proceeded by semicolon
       mov     5,_B0  ; for assembly instead of the single quote
                      ; for PicBasic
endasm
CLEARWDT              ' reset watchdog timer to make sure it is
                     ' reset prior to moving on
pause 100             ' delay 100 milliseconds
```

COUNT *pin,period,var*

This command is carried over from the Basic Stamp II. I like the simplicity of it. You can use it to sample a port for a specified amount of time, and then use the data to calculate frequency or speed or just the rate that pulses are arriving. The resolution of the sample time is dependent on the crystal frequency and it counts the transitions from low to high. At a 4-MHz crystal frequency the designated pin is checked every 20 microsecond. With a 20-MHz crystal/resonator it's checked every 4 microsecond.

Pin can be 0–15 or a variable with the value of 0–15 or a portname.number such as PORTA.1. The pin is automatically made an input. *Period* is in units of milliseconds. It can run up to 65536 milliseconds or 65.5 seconds but that's quite a long sample time. *Var* stores the total number of pulses counted during the *period*. It really assumes a 50% duty cycle that is equal high and low time but it will work with more or less than 50%. You have to know your signal to use this command accurately.

Example:

```
'Speed calculator assuming each pulse is 1/3600 of a mile and
'measure pulses for 1 second,
'(3600 second/hr)*(1 mile/3600 pulses)*(X pulses/second) = Y
'mile/hr or 1 pulse per MPH

distance        var     byte
speed           var     byte

sample:
count 0, 1000, distance          'get number of pulses in a
                                 'second
speed = distance                 ' number of pulses = MPH
serout 1, N9600, speed           ' send speed value serial to
                                 ' display device
goto sample
```

DATA *@location, constant, constant, ...*

This command is different from most of the other commands because it is only executed when the PIC is programmed, not when the program in the PIC is run. It is used to preload the EEPROM memory. It only works with PICs that have EEPROM memory, like the 16F84. The *location* variable is optional and sets the starting point to store values. If it is not included in the command then the location 0 is used. The *constant* value can be a numeric or string constant. Only the least significant byte is stored for numeric constants unless it is preceded by the modifier "word". If word precedes it, then the next two bytes are stored in consecutive EEPROM locations. String constants are treated as separate ASCII values and stored as separate values.

This command is similar to the PBC EEPROM command, and in fact PBPro allows the EEPROM command as well. They do the same thing—store constants in EEPROM at compile time—but they have slightly different formats. DATA doesn't require a parenthesis around the constants, but does require an @ symbol before the location value. DATA also has the 'word' operator and EEPROM doesn't. The command DATA was carried over from the BASIC Stamp 2.

Example:

```
DATA @5, 10,"A"           'Store value 10 at location 5, ASCII
                          'value of "A"
                          'at location 6 of the internal EEPROM

DATA   word $1234         'Stores $34 at first two bytes and
                          '$12 at the second 'two bytes.
DATA   $1234              'Stores $4 at first byte, $3 at
                          'second byte, $2 third 'and $1 fourth
```

DEBUG *item, item, ...*

This command was written to give PBPro users a simple way of monitoring the program while running in a PIC. It serially sends any variable data out a predefined port pin at a predefined baud rate in 8N1 format. It really is a simplified SEROUT command. It can be handy for monitoring a set of variables by connecting a PC or serial LCD module to display the values sent. Sometimes just seeing the value of a variable can clue you into why your program isn't working like you expected.

Because the DEBUG command uses predefined definitions, the only way to change those definitions is with a set of DEFINE statements. These statements are read when the PBPro program is compiled. They will stay intact throughout the program. That means all DEBUG commands inserted in a PBPro program will communicate at the same baud rate, via the same port pin. That is how this command differs from the SEROUT command. The SEROUT command sets all the communication parameters at each SEROUT command so each SEROUT command can drive a different pin at various baud rates.

DEBUG definitions include the following:

```
DEFINE DEBUG_REG  PORTB        'Set debug port to B
DEFINE DEBUG_BIT 0             'Use PORTB bit 0
DEFINE DEBUG_BAUD 2400         'Set baud rate to 2400
DEFINE DEBUG_MODE 1            '0 = true, 1 = inverted
DEFINE DEBUG_PACING 1000       'This sets the delay between
                               'characters its in 1 msec
                               'units so 1000 = 1 second
```

DEBUG can be enabled or disabled with the DISABLE DEBUG and ENABLE DEBUG commands.

DEBUGIN *{timeout, label,} [item, item, ...]*

This command was written to expand on the capabilities the DEBUG command offers. The DEBUG command focused on sending information out. This command is for receiving information in. It really is just another form of SERIN2 command, explained later, but it uses less code space than SERIN2. The real advantage to this command is the ability to build a debugger program that will stop the program and wait for you to input data. Then the program can run from that point on. DEBUGIN can be enabled or disabled with the DISABLE DEBUG and ENABLE DEBUG commands.

DISABLE INTERRUPT

This is one of the unique PBPro commands that came about because of interrupts. The biggest complaint I have always had with BASIC Stamps, and even PBC, was the lack of interrupts in BASIC. In fact, the BASIC Stamps don't offer interrupts at all. Interrupts add so much to a program. You can have the program interrupt when a switch is closed or an internal timer runs out, and then run another short program called an *interrupt handler.*

When the interrupt handler is done, the PIC is returned to the main program loop where it was interrupted. This allows functions to be monitored in the background while your main program loop is running. That is the great advantage of interrupts.

Sometimes during your main loop you may be doing something that an interrupt could screw up. That's when the DISABLE INTERRUPT command is used. Place it before the code you don't want interrupted and the interrupt handler routine will be disabled. The ENABLE INTERRUPT command mentioned later will restore the interrupts. Using DISABLE INTERRUPT does not mean the interrupt handler routine won't ever function, its just delayed until the ENABLE INTERRUPT command is encountered.

```
DISABLE INTERRUPT      ' Disable interrupts
For x = 1 to 3         ' Sample PORTB three consecutive
value(x) = portb      ' times without interruption.
next
ENABLE INTERRUPT       ' Ready for interrupts
```

DISABLE DEBUG

This command does a similar function as DISABLE INTERRUPT above, except it only works on the DEBUG commands. ENABLE DEBUG reverses this command.

DISABLE

This combines the DISABLE DEBUG and DISABLE INTERRUPT into one command.

It disables both the DEBUG and ON INTERRUPT functions. ENABLE command resets them back to working mode.

DTMFOUT *pin, onms, offms, [Tone,Tone, ...]*

This is a clever command that generates the Touch Tones your phone uses to call a number. *Pin* designates which PIC pin to send the signal out of. It can be the 0–15 value or a variable or the portname.number format (PORTA.0). *Onms* is the number of milliseconds each tone is on. *Offms* is the time delay in milliseconds between tones. These are optional parameters to include because the command defaults to an *onms* of 200 msec and an *offms* of 50 msec.

The *Tone(s)* are a number from 0 to 15. The digits 0 to 9 are the same tones as the numbers on your Touch Tone phone. The * key tone is number 10 and the # key is number 11. Values 12–15 are reserved for the extended keys A, B, C and D, respectively. Extended keypads offer those keys.

This command actually uses another PBPro command, FREQOUT, which I'll discuss later. The DTMFOUT uses FREQOUT to produce two tones in a pulse-width modulated method. The actual signal is not very useful until you add external wave shaping and, for best results, an amplifier to boost the sound. This command also prefers you run the PIC at 20 MHz rather than the default 4 MHz for best results. A suggested filter is shown in Figure 3-2.

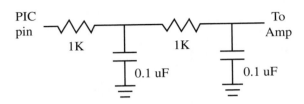

Figure 3-2: Output filter for use with tones generated by the DTMFOUT command.

EEPROM *location, (constant, constant)*

This command is the same as the PBC version except it uses '[ ]' instead of '( )'. It is only executed when the PIC is programmed, not when the program in the PIC is run. It is used to preload the EEPROM memory. It only works with PICs that have EEPROM memory like the 16F84. The *location* variable is optional and sets the starting point to store values. If it is not included in the command then the location 0 is used. The *constant* value can be a numeric or string constant. Only the least significant byte is stored for numeric constants and string constants are treated as separate ASCII values. This command is similar to the DATA command mentioned earlier and is included mainly for people converting Basic Stamp 1 code to PBPro code.

Example:

```
EEPROM 5, (10,"A")    'Store value 10 at location 5, ASCII value
                      'of "A" at location 6 of the internal EEPROM
```

ENABLE INTERRUPT

This command enables the interrupts that were put on hold by the DISABLE INTERRUPT command mentioned above. Both ENABLE INTERRUPT and DISABLE INTERRUPT take up zero code space and are really directives to the compiler to include some interrupt jumps or not. ENABLE INTERRUPT restores the interrupt handler to run mode if an interrupt occurred during the DISABLE INTERRUPT mode.

If an interrupt occurred, the interrupt handler will start immediately after the ENABLE INTERRUPT command and before the first PBPro command that follows ENABLE INTERRUPT. DISABLE INTERRUPT and ENABLE INTERRUPT should always surround the interrupt handler routine. You don't want to receive an interrupt while your processing a previous interrupt or you could end up never leaving that section of code.

Example:

```
DISABLE INTERRUPT              ' No interrupts please!
inthand:
      total = total + 1        'increment interrupt count for
                               'whatever purpose
      resume                   'Return back to the main loop and
                               'reset the interrupt flag
ENABLE INTERRUPT               ' Interrupts enabled
```

ENABLE DEBUG

This command does a similar function as ENABLE INTERRUPT above, except it only works on the DEBUG commands. DISABLE DEBUG reverses this command.

ENABLE

This combines the ENABLE DEBUG and ENABLE INTERRUPT into one command.

It enables both the DEBUG and ON INTERRUPT functions. DISABLE command resets them back to working mode.

END

This command stops program execution. It is automatically placed if you don't include it but it's best to include it at the end of your main loop. Place the subroutines after this command. You should never actually get to this command since your PIC stops and goes into an endless series of NAP commands and never leaves until you reset the PIC.

FOR ... NEXT

This command is familiar to anyone who has used BASIC. The format is as follows:

FOR *variable = start* TO *end [* STEP *[-] increment]*
        [Picbasic Routine]
NEXT {*variable*}

The PicBasic routine trapped between the FOR/NEXT command structure will be executed while the logical statement following the FOR command is within the *start* and *end* values.

Variable can be any variable you create with VAR mentioned earlier. *Start* and *end* are limited to the size of the *variable*. If the *variable* is a byte, then *start* and *end* must be 255 or less. If *variable* is a word size, then *start* and *end* must be less than 65536. What this command really does is first initialize the *variable* to the *start* value. It then executes the PicBasic Routine. At the end of the routine the *variable* is incremented by one and compared to the *end* value. If *variable* is greater than the *end* value, then the PicBasic command that appears after the NEXT command is executed. If *variable* is less than or equal to the *end* value, then the trapped PicBasic Routine is executed again.

The STEP option allows the command to do something other than increment the *variable* by one. It will instead increment the *variable* by the value *increment*. If *increment* is a negative number, then the *variable* is actually decremented. If a negative number is used, you must make sure *start* is a greater number than *end*.

The *variable* name after NEXT is optional. It will increment the closest FOR *variable*. If you have a FOR ... NEXT loop within a FOR ... NEXT loop, then it's best to place the proper *variable* name after the NEXT.

Here is an example of FOR ... NEXT used to count in binary with LEDs connected to PORTB:

Example:

```
X       var       byte

FOR X = 0 to 255                'Count from 0 to 255
PORTB =   X                     'LED anodes connected to PORTB, cathodes
                                'ground
NEXT                              'loop until X >255
```

FREQOUT *pin, onms, frequency1, frequency2*

Just like the DTMFOUT command this command produces a pulsed output of one or two different frequencies. It also needs the same filtering and amplification that DTMFOUT needed for sound, unless you just want a square wave. FREQOUT actually produces a square wave on the designated *pin*. It is different than DTMFOUT in that you define the output frequency by setting a value of 0–32767 in hertz for each *frequency* variable. *Frequency1* is necessary, but *frequency2* is optional.

Pin is the port you want to send the signal out from. As usual it can be 0–15, a variable or the portname.number (PORTB.1) option. This command only lets you define the *onms* "on time" in milliseconds. FREQOUT is very handy for driving a piezo buzzer. Buzzers can be used to indicate time has run out or give audible feedback that a switch was pressed.

Example:

```
' Timer example with buzzer output

x        var      byte

for x = 0 to 60                  'not very accurate 1 minute timer
pause 1000                       '1 second pause repeated 60 times
next                             'loop if x < 60

freqout 0, 1000, 2000            'output a square wave for 1 second at
                                 '2000 hertz
                                 'indicating 1 minute has elapsed
```

GOSUB *label*

This command could be considered a temporary GOTO. Just like GOTO, it jumps to the defined label. Unlike GOTO, the program will return back and continue with the command after GOSUB when the RETURN command is encountered. GOSUB is really an abbreviation for GOto SUBroutine. A subroutine is a program listing within a main program. You can have several subroutines that each performs a special function. You can also place a common routine in one subroutine rather than write the common routine multiple times. This is a way to save memory.

You can also GOSUB within a subroutine. The first return will bring you back to the subroutine and the second return will bring you back to the original GOSUB. This is known as nesting. You are limited to four levels of nesting with PBPro or, in other words, a maximum of four GOSUB commands may be used together. The return is performed by an accompanying command RETURN. They must both be in the program to make the function work.

You can have multiple GOSUB commands jumping to the same routine but only have one RETURN command at the end of the subroutine. This is quite common. The total number of GOSUB commands is unlimited—just don't nest more than four.

Example:

```
FLASH:
GOSUB   SUB              ' Jump to subroutine SUB
        GOTO   FLASH     ' Loop again to flash LED on PORTB bit 4

SUB:
        TOGGLE   4       ' Change state of PORTB bit 4
        RETURN
```

GOTO *label*

This command redirects the current program location to a new location. This can be used for bypassing a section of code accessed by another part of the program or even jumping back to the start of the program. *Label* must be defined somewhere else in the program. It is usually placed at the bottom of the program just above the END command. That way you can direct the program back to the beginning or initialization before reaching the END command.

Example:

```
GOTO      START    ' Jump to the beginning of the program at label
                   ' START
END
```

HIGH *pin*

This command sets a specific bit in one of the predefined PIC I/O data registers and then clears the corresponding TRIS register bit to 0 to make it an output. The predefined PIC I/O data and TRIS register are dependent on the PIC you are programming. The I/O control section earlier in this chapter defines the ports used. The *pin* value designates which port bit to operate on. *Pin* must be a number from 0 to 15.

This command was handy in PBC, but PBPro makes it so much easier with the portname.number format that HIGH is rarely needed and takes up more code space. If you like using it, PBPro still includes it to maintain compatibility with PBC and BASIC Stamps.

Example:

```
HIGH    1          'Set PORTB bit 1 high and make it an output. (PIC
                   'pin 7 on 16F84)
```

alternative,

```
TRISB.1 = 0               'make PortB bit 1 an output
PORTB.1 = 1               'make PortB bit1 high
```

HSERIN *paritylabel, timeout, label, [item, ...]*
HSEROUT *[item, ...]*

These are some of the commands that put PBPro in a league of its own. Some of the PICs have a built-in serial port peripheral. That peripheral is used to do serial RS232 type communication and offers many features that bit banging a RS232 signal doesn't offer. To set up the peripheral requires several steps but this command takes care of it for you. The formats are similar to the SERIN/SEROUT commands except the baud rate and other transmit set-up details are preset using a DEFINE statement. These DEFINE statements are used when the program is compiled.

The HSERIN command offers several options in the command line such as *paritylabel, timeout, label*. The only required operator is the list of *items*, which is the list of variables where you want the received information to be stored. After that is set, you can add the optional *timeout* value in milliseconds that specifies how long to wait for sent information before jumping to the *label*. Use the *timeout* option to prevent the program from locking up at this command while it waits for information. In the DEFINE set-up of this command you can set-up various parity options. *Paritylabel* is used as the place to jump to if there is a parity error.

The DEFINE statements include:

```
DEFINE   HSER_RCSTA 90h                    'enable receive register in
                                           'PIC
DEFINE   HSER_TXSTA 20h                    'enable transmit register in
                                           'PIC
```

```
DEFINE   HSER_BAUD 2400          'set baud rate
DEFINE   HSER_EVEN  1            'set even parity
DEFINE   HSER_ODD   1            'set odd parity
```

Setting HSER_EVEN and HSER_ODD to 0 will disable the parity mode.

The HSEROUT command line requires a list of *items* just like the HSERIN command but this time it's the list of characters to send. There are no optional parameters in the command line for HSEROUT except for the data modifiers described briefly below.

Both HSERIN and HSEROUT commands allow you to place modifiers on the information being sent and received. The PBPro manual explains these fairly well so I'll just touch on them here.

Bin, dec or hex may precede the variable being sent or received. These will convert the value of the variable into binary, decimal or hexadecimal prior to sending it or if received, prior to storing the value. For example:

```
HSEROUT  [dec b0, 13] 'send value of b0 as decimal and then 13
                      'which is ascii carriage return.
                      'If b0 = 123 then "1" then "2" then "3"
                      'would be sent
```

If the "dec" modifier wasn't included in the example above, then the contents of b0 would be sent as a byte value "123".

For HSEROUT you can also issue the REP or repeat modifier. It will send a character as many times as you specify. The format is variable/number of sends.

For example:

```
HSEROUT  ["a"/4]                     'send "a" 4 times or "aaaa"
```

I2CREAD *datapin, clockpin, control, address, [var,var,...], label*
I2CWRITE *datapin, clockpin, control, address, [value, value, ...], label*

If you ever want to communicate using the Phillips I^{2}C format, these commands are designed to make it easier for you. The *datapin* is the PIC pin you want to communicate on and is the PIC pin connected to the data pin of the I^{2}C device. The *clock-*

pin is the PIC pin you want to send the clock pulse on and is the PIC pin connected to the clock pin of the I²C device. The *control* is the code for the I²C device that communicates chip select info, address info, and other things that the I²C device defines. This control code can be found in the data sheet of the I²C device you are using. Even the PBPro manual has some codes for some of the Microchip EEPROM parts. Getting the control character wrong is a very common problem, so double-check it.

The *address* is the location the information is to be read from or written to. Typically this will be a variable that you change somewhere else in your PBPro program. The address can be a byte size or word size, but it is driven by the I²C device you are using. Again, check with the data sheet for the I²C device.

The *var* in the I2CREAD command is the list of variables where you store what you've read. If a word-sized variable is used to store the data, the high order byte is filled first and then the low order byte. Remember this because it's opposite most other two-byte operations in PBPro.

The *value* in the I2CWRITE command is the list of data to write into the I²C device. If *value* is a word-sized variable, then the high byte is sent first followed by the low byte. The *var* or *value* data must be contained within brackets, not parentheses.

The *label* is optional but handy. If the I²C device does not send an acknowledge signal to the PIC, then the PBPro program will jump to the label. This is handy for error detection. Your PBPro program can have a routine written at *label* to either warn you somehow or have the program resend the data.

Here is an example of I²C communication. I strongly recommend you consult the PBPro manual on this command because it is written quite well.

```
' I2CREAD and I2WRITE Commands
'
' Write to the first 10 locations of an external 24LC01B 128 byte
' serial EEPROM
' The control code is %1010xxx0 (xxx means don't care) or $A0
' Then read first 10 locations back and send them serially to a PC


SO        con      0               ' Define serial output pin
DPIN      var      PORTA.0         ' I2C data pin
```

```
CPIN      var      PORTA.1                          ' I2C clock pin
x         var      byte
y         var      byte
z         var      byte

init:
y=0

main:
for x = 0 To 9                          ' Loop 10 times
y = y + 2
I2CWRITE        DPIN,CPIN,$A0,x,[y]      ' Write the value of y at each
x address
pause 10                                ' Delay 10ms after each write
next

loop:
for x = 0 To 9                          ' Loop 10 times
I2CREAD         DPIN,CPIN,$A0,x,[z]      ' Read the stored values
serout SO,N2400,[x, z, 10, 13]          ' Print location, value then
next line and carriage return

communication.
 next                                   ' on PC screen via RS232

end
```

IF..THEN..ELSE..ENDIF
IF *comp* AND/OR *comp* THEN
        statement1
ELSE
        statement2
ENDIF

If you're at all familiar with the BASIC language, you will recognize this command. PBPro expands on the simple IF..THEN command of PBC by adding the ELSE option. The ELSE option gives another alternate to the original IF .. THEN statement. If the comparison is true, then do statement 1. If the comparison is not true, then do statement 2. The IF .. THEN format works the same as PBC if the ELSE/ENDIF lines are left off.

Example:

```
X        var      byte
LED      var      portb.0          'LED is connected to portb pin 0

init:
         LED = 0
             x = 0
main:
x = x + 1                          'increment x
if x = 100 then                    'test x value
         led = 1                   'LED on.
         x = 0                     'clear x
         pause 1000               'wait 1 second so a human can see the
                                   'LED is on

         goto main
else
         led = 0                   'LED off
         goto main
```

INPUT *pin*

This command is carried over from PBC to make a pin an input. It works on the first 16 I/O pins defined for a PIC, as mentioned in the I/O section of this chapter.

The *pin* variable is the I/O pin (0-16) you want to modify the direction of. *pin* can also be a variable, which is handy. The command INPUT 2 makes PORTB pin 2 an input. The limitation is the 16 I/O. If you are using a PIC with more than 16 I/O pins, then you have to set the I/O direction by modifying the TRIS register for that PORT. It also takes less assembly code in the compiler to do the TRIS modification then the INPUT command. For those two reasons its best to not pick up the habit of using the INPUT command.

To show you the difference between INPUT and TRIS, I'll show the two methods with the same results.

Example:

```
INPUT 2          'Set portb pin 2 to an input
```

or

```
TRISB.2 = 1              'Set portb pin 2 to an input.
```

{LET}

This command is in brackets because you never actually need to use it! When you define an expression, the LET command is implied. For example:

X= X + 1 is the same as LET X = X + 1

So why did I mention this command? Just to let you know the compiler will allow you to use LET if you really want to! But since there's no need to, why bother?

LCDIN *{address,} [var, var, ...]*

This command was created after the LCDOUT command mentioned next. Some LCD modules have extra RAM space on-board, and this command allows you to access that space for extra storage.

You really need to study the LCD data sheets to understand this function. It can be a handy command, but I have never had a need for it. I find the larger EEPROM PICs offer me all the extra storage I could ever need.

LCDOUT *item, item, ...*

This command is one of the neat new features added in PBPro. Many people have used expensive add-on modules that convert a parallel drive LCD to a serial interface. The original idea of the add-on modules was to save I/O, but people often used them just because it made driving the LCD easier. This PBPro command practically eliminates the need for the converter module if you have the seven I/O pins required to drive the LCD because it reduces the software to drive the LCD down to one single command. The command is written to drive an LCD module built around the Hitachi 44780 controller (which 99% of them are). PBPro defaults to a predefined I/O set-up to connect to the LCD. That set-up can be changed using the DEFINE statement.

PBPro defaults to using PORTA.0 - PORTA.3 for the DB4 - DB7 data pin connections. It uses PORTA.4 for the RS (*Register Select*) connection and it uses PORTB.3 for the E (*enable*) connection. It also sets the LCD up for 4-bit buss communication using DB4-DB7 and 2-line display. These can all be changed with the following DEFINE statements:

```
LCD_DREG                'Port used to connect to DB4-DB7
LCD_BITS                'Buss size used in the LCD (4 or 8 bits)
LCD_DBIT                'First port pin to be connected to DB4-DB7
                        ' (4-bit) or DB0-DB7 (8-bit)
```

Example:

```
'Change the LCD data buss to 8 bits, connected to PORTC, using all
'8 PORTC pins
DEFINE       LCD_BITS      8                  'Buss set to 8 bits
DEFINELCD_DREG        PORTC    'Use PORTC
DEFINELCD_DBIT        0                 'PORTC pin 0 is the first data
                                        ' pin connected to DB0 of LCD.
```

```
LCD_RSREG               'Port used to connect to the LCD RS pin
LCD_RSBIT               'Port pin to be connected to the LCD RS pin
LCD_EREG                'Port used to connect to the LCD E pin
LCD_EBIT                'Port pin to be connected to the LCD E pin
```

Here's another example:

```
'Make port C bits 0,1 the RS and E connections
DEFINELCD_RSREG       PORTC    'Use PORTC for RS
DEFINELCD_EREG        PORTC    'Use PORTC for E
DEFINELCD_RSBIT       0        'Connect LCD RS to pin 0
DEFINELCD_EBIT        1         'Connect LCD E to pin 1
```

The next DEFINE statement is the one you may use the most. With it you set the number of lines the LCD contains. A 2x16 LCD, which is common, has two lines of 16 characters. Another common LCD is the 4x20 size that has four lines of 20 characters. This has to be set up with a DEFINE to allow the LCDOUT command to work properly:

```
DEFINE LCD_LINES   4            'set the number of lines to 4 for 4x20
LCD
```

Once you have the LCD set-up defined, you can start sending characters to it. All characters sent are the value that the LCD has associated to each character. Most of the common characters, such as numbers and letters are associated with the

ASCII value for them. Other special characters have to be looked up in the LCD data sheet. Because the ASCII values are used for common characters, the LCDOUT command allows you to use the same modifiers used in the SEROUT/SERIN commands.

The # symbol preceding a variable will the send the ASCII value for each number in the variable. For example, if the variable B0 = 255, then three bytes would be sent ASCII for "2", "5", "5" if the # was used. Otherwise, the value $FF (255) would be sent if the # sign was omitted. Another shortcut is the sending of strings. If the characters to be sent are within a set of quotes, then each character's ASCII value is sent. Here are examples of both:

```
LCDOUT "Test of LCD"          'The characters "T", "e", "s", … ,
                              '"D" are sent
LCDOUT #B1                    ' If B1 = 128 then "1", "2", "8" is
                              ' sent
```

The LCD module also can be controlled to do certain functions like return the cursor to home or clear the display. They are necessary to format the LCD display the way you want. The LCD data sheet will explain these more, but below are several common commands. The LCDOUT command allows you to send the control commands right in the LCDOUT command line. The code $FE is sent first and the next byte sent is the command. The list of common commands is below.

$01   Clear display

$02   Return home

$0C   Cursor off

$0E   Underline cursor on

$0F   Blinking cursor on

$10   Move cursor left one position

$14   Move cursor right one position

$C0   Move cursor to beginning of second line

Here's an example:

```
'Display the value of B0 in numeric form on an LCD. The display
'will count from 0 to 255 and 'then rollover to 0 again.
```

```
B0      var     byte
loop:

        LCDOUT $FE, 1, #B0    'clear the LCD and then display
                             'the value of B0
     B0 = B0 + 1             'increment b0 by 1
     pause 1000              'delay for 1 second
     goto loop               'loop again
```

LOOKDOWN *search,[constant {, constant}], var*

It can be difficult to remember what this command does; I look it up in the manual almost every time I use it. What it does is look down a list of values (*constant*) and compares each value to a master value (*search*). If a match is found, then the position of the *constant* is stored in a variable (*var*). It is a look-up table method for converting any character into a numeric value from 0 to 255. If *search* matches the first *constant*, then *var* is set to 0. If the second *constant* matches *search,* then *var* is set to 1, etc. String constants and numeric constants can both be part of the table.

PBPro works the same way as PBC does. It separates the list of constants by looking at each 8-bit value. It's best to separate the constants with commas so the compiler knows where to start and where to stop—1010 is not treated the same as 10,10. If you use string constants, then they will be treated as their respective 8-bit value. Therefore commas may not be needed for string variables. The only real difference between PBC and PBPro is this command uses brackets instead of parentheses.

Here's an example:

```
LOOKDOWN   B0,[0, 1, 2, 4, 8, 16, 32, 64, 128], B1    ' B1 contains
                                                      ' in decimal
                                                      ' which
                                                      ' single bit
                                                      ' is set
                                                      ' in B0. If
                                                      ' B0 = 128 or
                                                      ' 10000000
                                                      ' binary then
                                                      ' B1 = 8.
' If more than one
' bit is set or no bits are set in 'B0 then B1 = 0
```

LOOKDOWN2 *search,{test}[constant {, constant}], var*

This command works the same as LOOKDOWN, but it has three advantages over it.

For one, this command takes care of the page boundary problem in the PIC. If your program runs over the first 2k of memory space, then this is the command to use. The LOOKDOWN command uses less code space then LOOKDOWN2, but if the table of constants crosses the page boundary then the LOOKDOWN command will not work properly. PBPro will warn you of page boundary crossings when it compiles the program, but you still have to be aware of which command to use.

A second advantage of LOOKDOWN2 is it allows 16-bit constants. The result is still a byte, since only 256 constants are allowed. However, the *search* parameter can now be 16 bits long.

Finally, LOOKDOWN2 also adds the optional *test* parameter. This parameter can do a logical search rather than just a search for the constant that equals the *search* value. For example, using a > or < symbol will make the command look for the first constant greater than or less than the *search* value.

Example:

```
Lookdown2 W0,< [100, 200, 300, 400, 500], B0      'Look for the
                                                  'first constant
                                                  'less than W0
                                                  'and return the
                                                  'position in B0
```

LOOKUP *index,[constant {, constant}], variable*

This command performs a lookup table function. *Index* is an 8-bit variable that is used to choose a value from the list of *constants*. The selected *constant* is then stored in the *variable* following the list of *constants*. If the *index* variable is 0, the first constant is stored in the *variable*. If index is 1, then the second *constant* is stored in the *variable,* and so on. If *index* is a value larger than the number of listed constants, then variable is left unchanged. The constants can be numeric or string constants. A comma should separate each constant. This command is the same as the PBC command except it uses brackets instead of parentheses.

Here is an example:

```
FOR B0 = 0 to 7                              'Convert decimal number to
LOOKUP  B0,[0, 1, 2, 4, 8, 16, 32, 64, 128], B1  ' a single bit to
                                             ' be set.
NEXT
```

LOOKUP2 *index,[constant {, constant}], variable*

This command works the same as LOOKUP but has two advantages. For one, this command takes care of the page boundary problem in the PIC. If your program runs beyond the first 2k of memory space, then this is the command to use. While the LOOKUP command uses less code space than LOOKUP2, if the table of constants crosses the page boundary the LOOKUP command will not work properly. PBPro will warn you of page boundary crossings when it compiles the program, but you still have to be aware of which command to use.

A second advantage of LOOKUP2 is it allows 16-bit constants. The *variable* must be word size if the *constant*s are 16 bits long. The *index* is limited to a byte size because only 256 constants are allowed.

Example:

```
Lookup2 B0, [100, 200, 300, 400, 500], W0 'Store the constant in
                                          'W0 that is pointed
'to by B0. If B0 = 0 then W0 = 100, B0=1,
'W0=200, etc.
```

LOW *pin*

This command clears a specific bit in one of the predefined PIC I/O data registers and then clears the corresponding bit in the TRIS register to make that pin an output. The predefined PIC I/O data and TRIS register depend on the PIC version you are programming. The I/O control section earlier in this chapter defines the ports used.

The *pin* value designates which data register and TRIS register bit to operate on. *Pin* must be a number from 0 to 15. This command was handy in PBC, but PBPro makes it so much easier with the portname.number format that LOW is rarely needed and just takes up more code space. If you like using it, PBPro still includes it to maintain compatibility with PBC and BASIC Stamps.

Example:

```
'Assume we are programming a 16F84 PIC which uses PORTB (0-7) and
'PORTA (8-12)
LOW   1         'Clear PORTB bit 1 and make it an output. (PIC pin
                '7 on 16F84)
```

or

```
TRISB.1 = 0            'make PortB bit 1 an output
PORTB.1 = 0            'make PortB bit1 low
```

NAP *period*

This command places the PIC in a low power mode for short periods of time. This can be used to save power in battery applications. Nap for a short time, then go check the I/O, then nap again. The *period* variable is a number from 0 to 7. Each corresponds to the delay listed below. The timing is derived from the Watchdog Timer inside the PIC. You must have the WDT turned on when you program the PIC to use this.

The WDT is driven by an internal RC circuit, so its accuracy is not great. All time values are nominal but can vary as much as 20%. Sometimes you will use one of the PIC's internal timers, which typically require you to modify the timer prescaler. If you play with the prescaler in the PIC, you can affect the timing of the NAP instruction.

period	delay (approx)
0	18 msec
1	36 msec
2	72 msec
3	144 msec
4	288 msec
5	576 msec
6	1.152 sec
7	2.304 msec

Here's an example:

```
loop:
        if portb.0 = 0 then prog    ' Test portb pin 0 if its low
        nap 6                        ' low power for 1.152 sec
                                     ' delay
        goto loop                    ' test pin 0 again

prog:
        toggle portb.1               ' portb pin 0 low,  toggle
                                     ' portb pin 1
        goto loop                    ' test pin 0 again
```

ON DEBUG GOTO *label*

This single command allows your PBPro program to help you monitor your program's operation internally. With this command, you can jump to a DEBUG routine at the location *label*. It can be another PBPro routine that sends out the current variable values. With a little bit of work on your PC, you could make a emulator type program to read your variables and display what is going on after every PBPro command. This is because this command inserts a jump to your debug monitor routine before each PBPro command.

This is really an advanced step and really requires deeper explanation than we give it here, so you should consult the PBPro manual before attempting to use it. Unlike the ON INTERRUPT command discussed next, this command runs right within the normal program space and doesn't monitor any PIC registers to determine if a jump should occur. It was developed to allow in-process monitoring of your PBPro code.

To use this command, you need to establish a word-size system variable named DEBUG_ADDRESS. This is required so PBPro has a place to store the address where the program was interrupted. When the DEBUG routine is done, it uses that location to return to your main program.

Here's an example:

```
'Send out status of B0 during program run time

B0      var     byte
DEBUG_ADDRESS   VAR     word
```

```
ON DEBUG GOTO dbuggr
loop:
b0 = b0 + 1                      ' increment b0
goto loop                        ' loop around again


dbuggr:
disable debug                    ' stop the debugger
LCDOUT $FE, 1, #B0               ' Display B0 on LCD
Return                           ' return to where we came from
enable debug                     ' enable debug after this routine
```

ON INTERRUPT   GOTO *label*

This single command makes PBPro worth its weight in gold. With this command, you can use the interrupt structure of the PIC without a bunch of assembly code. Interrupts can add a whole new dimension to a PicBasic program. If you're not familiar with interrupts, I'll explain them here briefly but I would also recommend you read the PIC data book section on interrupts to really understand how they work in a PIC.

Interrupts do just what the name implies: interrupt the PIC program. In the hardware of the PIC is circuitry that will stop execution of the main program and jump to a designated location. That location is a separate program called the *interrupt handler*. It is typically a very short section of code. When an interrupt occurs the interrupt handler runs and, when it's completed, the PIC gives control back to the main program right where it was interrupted. The main program then continues.

The PicBasic command ON INTERRUPT allows you to interrupt a PicBasic program and have the interrupt handler also written in PicBasic. The real advantage to using interrupts is the interrupt handler acts like a second program running in parallel with the main program, but all contained in a single PIC. In fact, the PIC offers several different ways to interrupt the main program.

Each different interrupt has its own flag that gets set in the INTCON register. If your interrupt handler first checks which flag is set, then the interrupt handler can run a separate program for each interrupt type. In this way multiple programs are running in parallel within a single PIC. As I mentioned, the interrupt handler is typically very short because the main program is briefly stalled while the interrupt handler is running. This could result in the main program missing any inputs or outputs it may have to be updated. Some of the interrupts are timer overflow, change on portb, portb.0 state change, serial communication, PWM control, and more. This topic could be a chapter in itself and I can't cover it all here. What this section will do is show the key components to writing an interrupt in PicBasic.

The *label* in the command line is the label of the interrupt handler. When a interrupt happens, the *label* is where the PicBasic program jumps to. The interrupt handler looks like any PicBasic section of code but ends with the RESUME command. That command (described later in this chapter) is what sends control back to the main program. Two other commands work with ONINTERRUPT: DISABLE, and ENABLE. DISABLE turns off all interrupts and ENABLE turns interrupts on. Sometimes you will be running a section of your main program that you don't want interrupted until you're done. Before that section of code, place the DISABLE command. At the end of that section of code, place the ENABLE command to allow interrupts to occur. If you had an interrupt request while the interrupts were disabled, then the interrupt will occur immediately after the ENABLE command.

The ON INTERRUPT command is placed at the top of the program with the interrupt handler label defined. Then the INTCON register within the PIC is set-up to define which interrupts to allow and which to ignore. (Again, to fully understand this command you should read the PIC data book.) The interrupt handler is usually placed after the end of the main program. This way it acts like a little subroutine that only gets run when called by an interrupt. DISABLE should be placed before the interrupt handler so a second interrupt doesn't interrupt it. Just think what would happen if the interrupt handler kept getting interrupted—the program would be stuck running the interrupt handler over and over again, never returning to the main program. At the end of the interrupt handler is the RESUME command and following that is the ENABLE command just in case additional subroutines are added later. Here is a simple interrupt in PicBasic example:

```
'Count the number of time RA4/TOCKI receives 256 pulses and
'display it on an LCD.

B0      var     byte

ON INTERRUPT GOTO inthand

INTCON = %10100000              'Enable the TMR0 timer overflow
                               'interrupt
```

```
OPTION_REG = %10100111        'Make TMR0 increment on the low to
                              'high
                                     'transition of the RA4/TOCKI
                                     'PIC pin.
main:
LCDOUT $FE, 1, #B0            'Display B0 (number of 256 pulses) on
                              'LCD

goto main

DISABLE
inthand:                             'TMR0 overflowed so 256 more
                                     'pulse were received
                                     'number of times 256 pulses
B0 = B0 + 1                          'were received

RESUME
ENABLE
```

OUTPUT *pin*

This makes a specific bit in one of the 16 predefined PIC ports an output. You must be careful to know what state the data register for the port is in before you issue this command. As soon as you issue this command, the status of the bit in the data register (high or low) will instantly show up at the PIC pin. Another way to do this is by directly modifying the TRIS register. Here are the two examples.

```
OUTPUT 1                     'Make PORTB bit 1 an output (PIC pin 7 on
                             '16F84)
```

or

```
TRISB.1 = 0                  'Make PORTB bit 1 an output
```

PAUSE *period*

This is a very useful command. It can pause program execution for a *period* of 1 to 65,535 milliseconds. It doesn't put the PIC in low power mode, but is more accurate than either the NAP or SLEEP command. You can use it to control timing or pulse widths or whatever your program requires. Here's an example:

```
trisb.0 = 0      'Make portb pin 0 an output
pulse:
        portb.0 = 1      'send high signal out pin 0
        pause 10         'pulse width is 10 msec
        portb.0 = 0      'send low pulse out pin 0
        pause 10         'pulse width is 10 msec
        goto pulse       'loop again to make square wave of close to
                         '50% duty cycle
```

PAUSEUS *period*

This command expands on the PAUSE command by allowing microsecond delays. The value *period* is still a number from 1 to 65,535, but it's in microseconds. Because the PIC internally runs a routine for every PicBasic command, this command is actually limited to how short the pause can go. At 4 MHz, the PIC can only accurately delay to a minimum of 24 microseconds. Values below this will not be accurate. If you use the DEFINE OSC operative to change the crystal frequency, then smaller delays can be obtained. Here are the minimum delay times:

4 MHz: 24 microsecond

8 MHz: 12 microsecond

10 MHz: 8 microsecond

12 MHz: 7 microsecond

16 MHz: 5 microsecond

20 MHz: 3 microsecond

Here is an example:

```
trisb.0 = 0              'Make Portb pin 0 an output
pulse:
        portb.0 = 1          'send high signal out pin 0
        pauseus 100          'pulse width is 100 usec
            portb.0 = 0          'send low pulse out pin 0
            pauseus 100          'pulse width is 100 usec
            goto pulse       'loop again to make square wave of close
                             'to 50% duty cycle
```

PEEK *address, var*
POKE *address, var*

These commands are listed as part of PBPro to carry over PBC programs. They should not be used in PBPro programs, however, since they don't work properly within the structure of PBPro.

PBPro allows you to directly access any register by its PicBasic name listed in the file PIC14EXT.bas. Those names should be used since they are much more efficient and much easier. As seen in previous examples, do the direct access rather than POKE and PEEK. For example,

```
INTCON = %10010000     'Set the interrupt control register
```

POT *pin, scale, var*

The POT command was developed to allow analog-to-digital measurement with a standard PIC I/O pin. Some PICs have built-in A/D ports that make the POT command unnecessary. Although an A/D port is far more accurate, you may want to use the POT command at some point.

In resistor and capacitor circuits, the rate of charge to reach a known voltage level in the capacitor is based on the values of the resistor and capacitor. If you know the charge time and the capacitor value, then you can figure out the resistance. That's how the POT command works. It uses the I/O pins' high and low thresholds as the trigger points for measuring the capacitor charging. The capacitor and resistor are connected to an I/O pin designated by the *pin* value. This command is the same as the PBC POT command except it can use predefined pins 0–15 or a variable that contains 0–15 or the portname.number format. A schematic is shown in Figure 3-3.

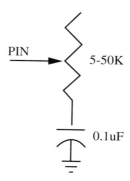

Figure 3-3: Circuit to be used with the POT command.

When the command is processed, the capacitor is first discharged by the I/O port, which is configured by the POT command as an output and low. After that, the I/O port is changed to an input and starts timing how long it takes for the capacitor to charge the high threshold voltage threshold of the PIC I/O port. At that point, the charge time is known and that time is converted into a 0–255 decimal value based on the value of the scale variable. 255 is the maximum resistance and 0 is minimum.

The key is the proper scale value that must be specified for this to work. In order to have the scale value match the resistance range, it must first be calculated for the R/C attached. No math is required because it must be determined experimentally. First set the resistance to its maximum value. Then set scale to 255 and run the command. The *var* value returned will be the proper scale value for that R/C combination.

Example:

```
POT    3, 240, B0      ' Measure the resistance and place the 0-255
                       ' value in B0 The 240 value was found first
                       ' by setting scale to 255
```

I don't recommend this command. PBPro makes programming A/D port so easy that I suggest you use a PIC with an A/D port. It is more useful and far more accurate.

PULSIN *pin, state, var*

This command allows you to measure the pulse width of an incoming square wave. It is great for doing pulse-width modulation (PWM) measurements. The resolution of this command is dependent on the oscillator used. If a 4-MHz oscillator is used, the resolution is 10 _sec. A 20-MHz oscillator has a 2 _sec resolution. The command will measure the high time if the *state* variable is a 1 or measure the low time if the *state* is 0.

The *pin* is the PIC pin to measure the pulse on. It can be the predefined port numbers 0–15, a variable containing 0–15, or the portname.pin method (PORTB.1). The measurement is stored in a variable defined by the *var* in the command line. It can be a word size or byte size variable. The command always measures 65535 samples but if the *var* is a byte size then the least significant byte is stored in the variable. If the signal is longer than 65535 samples, or no signal is present at all, then the *var* is set to 0.

The DEFINE OSC directive has no effect on this command. If a 4-MHz oscillator is used, then a measurement of 50 would mean a pulse width of 500 microsecond (50 * 10 microsecond resolution). If the signal is known to be a 50% duty cycle, then the total period of the signal is twice that measured or 1000 microsecond. Inverting that would give the frequency of 1 kHz. The example below uses pulsin to light a warning light if the frequency exceeds 1 kHz.

Example:

```
'Assuming a 50% duty cycle and 4mhz crystal, measure the high
'pulse width to 'determine the frequency.
'If the frequency is greater than 1 khz then light LED.

Pulse   var     word
LED     var     portb.1

measure:
pulsin portb.0, 1, pulse        ' Measure the high time of the
                                ' incoming pulse.
if pulse <= 50 then      ' Test sample if less than 1 khz
LED = 0                  ' LED off if <= 1 khz
goto measure                    ' Measure again
else
```

```
LED = 1                    ' Sample greater than 1 khz, light the LED
                           ' warning light.
goto measure
```

PULSOUT *pin, period*

This command is the complement to the PULSIN command in that it generates a pulse. The *pin* value determines the PIC pin the signal will come from. The pin value can be a predefined PIC pin number 0–15 or a variable with the value 0–15 or a portname.pin designator. The *period* value is actually the pulse width value. The *period* value determines how long the signal is high, and the resolution is determined by the oscillator. A 4-MHz oscillator will use a 10- microsecond resolution and a 20-MHz oscillator will have a 2- microsecond resolution. The *period* value can be a byte or word variable or a constant value from 1 − 65535. This command is commonly used to drive a servo, but can be used for various applications. You can use it to communicate with a shift register by using one pin for clock and the other for data. You can create a pulse-width modulated signal by placing the PULSOUT command in a loop. For example:

```
' Create a changing pulsewidth signal from 10 usec wide to 655
'msec wide.

Width    var    word

for width = 1 to 65535
pulsout portb.1, width
pause 1
next
```

PWM *pin, duty, cycle*

This command initially looks interesting because it implies it will create a variable pulse width like the PULSOUT example above. Instead, the command actually produces a series of very short pulses that are recreated a number of times. The resulting signal can be used "as is," but if a resistor-capacitor circuit is added then this command can be used to generate a voltage across the capacitor. By adjusting the *duty* value, the voltage across the capacitor can be varied. It creates a digital-to-analog (D/A) converter. The command does not run in the background, so it really is just for creating a burst. To drive a true PWM signal that continually runs in the background, you must use the PWM port available on some PICs.

To use this command *pin* is a predefined pin number 0–15, a variable with the value 0–15, or that portname.pin method of defining a pin. The *duty* value is the value that controls the number of short bursts put out. A value of 0 is 0%, or no pulses, and 255 is 100%, or a continuous pulse. The *cycle* value determines how long the burst of pulses repeats. The actual length of time for each cycle is determined by the oscillator used. A 4-MHz oscillator cycle time is about 5-msec long while a 20-MHz oscillator will result in a 1-msec time.

RANDOM *var*

This command produces a pseudo-random number for various applications. The *var* variable must be a word variable. It will produce a value from 1 to 65535, but will not produce zero. You cannot use a port number or port variable.

Example:

```
loop:
        random      W2          ' Create a random number
        pause 100               ' pause 100 msec
        goto loop               ' do it again
```

RCTIME *pin, state, var*

This command measures how long a pin stays at a specified level. It's another form of POT brought over from the Basic Stamp 2. It allows you to use a standard digital I/O port to measure resistance. I would rather use a PIC with an A/D port, but this is an alternative. The *pin* is the PIC pin your program is reading. The *state* is the level that is measured. The *var* is the variable that the time is stored in. The resolution of this command is based on the oscillator. A 4-MHz oscillator gives a 10-_sec resolution and 20 MHz gives a 2-_sec resolution. For example:

```
Time    var     word

portb.0 = 0             'set Port B pin 0 to low
RCTIME 0,0,time         'measure time pin 0 stays low
```

READ *address, var*

This command will only work on PIC devices with internal EEPROM like the 16F84. The *address* variable is the location to read. The value read at the specified location will be stored in the *var* variable. The command also has a special mode. If *address* is the value 255 then the total number of bytes available will be put in *var*.

Example:

```
'Subroutine

sub:
        read 10, b1              'Read value at location 10 in
                                 'variable b1
        return                   ' repeat
```

READCODE *address, var*
WRITECODE *address, var*

I'm breaking the alphabetical listing on these commands because they work so closely together. They are written to work with the 16F87X flash memory PICs. Those PICs allow you to read from and write over program code space. This opens so many possibilities, from simply adding more nonvolatile memory space (unused program memory) to creating a self-programming chip.

These are commands for the advanced user and are really beyond the scope of this book. Therefore, I'm covering them just to make you aware they exist. They operate similarly to the EEPROM commands, but they can be dangerous to your PBPro program because they can overwrite your code while the PIC is running. That would make the PIC lock up and then when the watchdog resets the PIC it would run till it locks up again—a very tough situation to diagnose. Here is an example of how these commands can be used:

```
value  VAR  word

loop:
writecode 100, value    ' Write the value variable content to memory
                        ' location 100
readcode 100, value          ' Read the word at location 100 and
                             ' store in value
goto loop                    ' loop forever
```

RESUME *{label}*

As mentioned earlier, this command is used at the end of an interrupt routine to direct the program back to where it was interrupted. RESUME also has the option of including a *label*. The program will jump to that label instead of back to the program. It can be useful if the interrupt routine is an error routine. That way the interrupt routine can possibly reset the program without the outside world ever knowing.

Example:

```
error:
        clear          ' reset all variables to 0
        resume start   ' go to the start label at the beginning of
                       ' the program
```

RETURN

This command is used at the end of a PicBasic subroutine to return to the command following the GOSUB command. For example:

```
Subrout:
        B0 = B0 + 1
        RETURN
```

REVERSE *pin*

This command changes the direction of the port or pin in the TRIS register. If a port was an output, it is changed to an input. If it was an input, then it's changed to an output. For example:

```
REVERSE 2               'Change direction of PORTB bit 2. (PIC pin
                        '8 on 16F84)
```

SERIN *pin, mode, timeout, label,[qual, qual], (#) item, item, ...*

This command emulates the RS232 communication common on PCs, also known as serial communication. With this command many interesting programs are possible. The command receives data from the sending source in 8N1 format that means eight data bits, no parity, and 1 stop bit. The *pin* variable is the PIC pin used. The *mode* variable is the baud rate to communicate at per the chart below.

This chart is slightly different than the BASIC Stamp because it allows 9600-baud communication in place of the Stamp's 600 baud. This is possible because a PIC programmed with PicBasic will run 15 times faster than a BASIC Stamp1. PBPro uses the same mode options as PBC, but you have to include a definition file if you want to use the mode names like T2400. That definition file is included with the command:

```
Include 'modedefs.bas"
```

Here are the *mode* options:

Mode value	Baud rate	Format
T2400 or 0	2400	TTL True
T1200 or 1	1200	TTL True
T9600 or 2	9600	TTL True
T300 or 3	300	TTL True
N2400 or 4	2400	TTL Inverted
N1200 or 5	1200	TTL Inverted
N9600 or 6	9600	TTL Inverted
N300 or 7	300	TTL Inverted

The *item* variable is the byte value received in the 8N1 format. If more than one *item* variable is listed in the command, then the program will wait for the exact number of *items* listed to be received. This can lock up a program while it waits for variables. In PBPro, this command has the *timeout* and *label* options to prevent that.

The *timeout* value is the amount of time in 1 millisecond units that the command will wait for a signal. If the signal does not arrive in the *timeout* time, then the program jumps to *label*.

The *qual* option is not needed but if used sets a prerequisite before accepting any *items*. The *qual* value can be a constant, variable or a string constant. The command looks for the *qual* to be received before going further.

The *item* variable can be preceded with a # character. This will convert any decimal number received into the ASCII equivalent and store that in the *item* variable. Any nondecimal values received will be ignored when using the #.

Example:

```
loop:
serin 1, n9600, 1000, error, ("A"), B0        ' Wait until the ASCII
' value for ' capital A is
' received or until 1 second '(1000 msec) has expired
' on portB pin 1 and then store 'the next byte in B0
        goto loop

error: high portb.1                            'Light LED on
                                               'PortB pin 1

        end
```

SERIN2 *datapin {\flowpin}, mode, {paritylabel}, {timeout, label}, [item, item, ...]*

This command was taken from the Basic Stamp 2 and extends the features of the SERIN command. I highly recommend that you read the PBPro manual on this command because it has so many features; I'll cover the basics here. The *mode* is set totally differently from the SERIN command. The *mode* contains more info than just the baud rate. It also selects the parity or no parity mode and it selects the signal style of inverted or true level. *Mode* is a 16-bit value but PBPro only uses the lower 15 bits.

The baud rate in mode is the first 0–12 bits. Bit 13 is the parity bit: 0- no parity and 1-even parity. Bit 14 is the signal style: 0-true and 1-inverted. Bit 15 is not used. The *mode* value is set by the equation:

$$\frac{(1000000 \,/\, baud) - 20}{}$$

$$+ 8192 \quad \textit{if even parity}$$

$$+ 16384 \quad \textit{if inverted}$$

---

mode

Here's an example:

To receive on port A pin 1, into a variable named `temp`:

300 baud, even parity, inverted

$(1000000 / 300) - 20 = 3313.3333$ or 3313

+ 8192   for even parity

+ 16384   for inverted

---

mode = 27889

```
SERIN2 porta.1, 27889, [temp]
```

SEROUT *pin, mode, [item, item, ...]*

This command sends a byte or bytes in serial 8N1 format out a specified pin. The *pin* variable sets the PORTB pin for communication. The *mode* value determines the communication baud rate. The chart below defines the mode options. Just like SERIN, to use the mode names such as T2400, you have to include the definition file:

```
Include "modedefs.bas"
```

Mode value	Baud rate	Format
T2400 or 0	2400	TTL True
T1200 or 1	1200	TTL True
T9600 or 2	9600	TTL True
T300 or 3	300	TTL True
N2400 or 4	2400	TTL Inverted
N1200 or 5	1200	TTL Inverted
N9600 or 6	9600	TTL Inverted
N300 or 7	300	TTL Inverted
OT2400 or 8	2400	Open Drain
OT1200 or 9	1200	Open Drain

OT9600 or 10	9600	Open Drain
OT300 or 11	300	Open Drain
N2400 or 12	2400	Open Collector
N1200 or 13	1200	Open Collector
N9600 or 14	9600	Open Collector
N300 or 15	300	Open Collector

The item *value*(s) can be in three formats and they can be mixed:

1) A string constant is sent as a string of characters i.e., "hello" is sent as five individual bytes.

2) A numeric value can be sent which will be received in a PC as the ASCII equivalent; i.e., 13 will represent the ASCII carriage return and 10 will be received as a line feed. If you send the numeric value to another PIC, though, it will be received as the binary value.

3) A numeric value preceded by a # symbol will break up the number and send it as individual ASCII characters. For example, #123 will be sent as "1", "2", "3". Here is how SEROUT is used:

```
loop:
        for b1 = 0 to 9           ' Send 10 numbers
        serout 5, n2400, [#b1, 10] ' 2400 baud inverted, send
                                  ' ASCII value of b1
        next                      ' followed by a line feed.
        goto loop
```

SEROUT2 *datapin {\flowpin}, mode, {pace}, {timeout, label}, [item, item, ...]*

This command comes from the Basic Stamp 2 format, so it operates similarly to the SERIN2 command. The *mode* is calculated similar to the SERIN2 command. Bits 0–12 set the baud rate. Bit 13 sets the parity even (bit 13 = 1) or no parity (bit 13 = 0). Bit 14 sets the true (bit 14 = 0) or inverted mode (bit 14 = 1). But SEROUT2 adds a bit 15 that sets the output to "always driven" (bit 15 = 0) or is left in high impedance open state (bit 15 = 1). The open state can be used to allow several modules on the same serial bus. The *mode* equation looks like this:

(1000000 / baud) - 20

  + 8192   if even parity

  + 16384   if inverted

  + 32768 if in open state

---

  mode

This command also adds an optional pace value that puts a delay in between characters. The pace value is in milliseconds and can range from 1 to 65535. I again recommend you read the PBPro manual on this command because it has so many features.

Here is an example of sending "hello world" out PORTB pin 3 at 300 baud, even parity, inverted, driven open with 10 msec between characters:

(1000000 / 300) - 20 = 3313

+ 8192 for even parity

+ 16384 for inverted

+ 32768 for open state

---

  mode = 60657

```
SEROUT2 portb.3, 60657, 10, ["hello world"]
```

SHIFTIN *datapin, clockpin, mode, [var {\bits}, var {\bits}, ...]*

This command allows your program to receive information a bit at a time and store it in *var*(s). You have for options to receive data that is set by the *mode* value.

0: First bit stored as most significant bit before sending clock pulse.

1: First bit stored and least significant bit before sending clock pulse.

2: First bit stored as most significant bit after sending clock pulse.

3: First bit stored and least significant bit after sending clock pulse.

The most confusing thing about this command is that the PIC generates the clock pulse. Normally the receiving chip gets the clock pulse from the sender. That's why these mode definitions are so important and initially confusing.

The *datapin* and *clockpin* define the PIC pins to use for these functions. The *bits* option allows less than 8 bits to be stored in a particular variable. If a variable was followed by a \7, then only 7 bits would be stored in that variable. For example:

```
Shiftin 1, 2, 0, [B0, B1\7]' Receive the data and store it LSB
                            ' first in B0 and
                            ' the next 7 bits in B1 LSB first
```

SHIFTOUT *datapin, clockpin, mode, [var {\bits}, var {\bits}, ...]*

This command allows your program to send a variable's information one bit at a time. You have two optional modes: least significant bit first (mode = 0) or most significant bit first (mode = 1).

If *bits* option is used, then only that number of bits are shifted out. The *datapin* and *clockpin* determine the PIC pins to use.

Example:

```
Shiftout portb.1, portb.2, 1, [B0, B1\7] 'Send the 8 bits of B0
 ' MSB first and the most significant 7 bits of B1
```

SLEEP *period*

This command places the PIC in low current mode and stops the PIC from running for a length of time. The *period* variable sets the amount of time to stay in the low-power mode. The SLEEP command timing is controlled by the Watchdog Timer within the PIC, which is just an RC circuit. To make the SLEEP timing more accurate it is compared to the system clock every 10 minutes and adjusted. The *period* value can range from 1 to 65535 in units of seconds. A value of 60 will make the PIC sleep for 60 seconds while a value of 65535 will make the PIC sleep for just over 18 hours.

For example:

```
loop:
        if portb.0 = 0 then prog    ' Test pin 0 if its low
        sleep 60                    ' low power for 1 minute delay
        goto loop                   ' test pin 0 again

prog:
        toggle 1                    ' pin 0 low,  toggle pin 1
        goto loop                   ' test pin 0 again
```

SOUND *pin,[note, duration {, note, duration}]*

This command was created to make sounds from a PIC. A PIC alone cannot produce sound so additional hardware is required as shown in Figure 3-4.

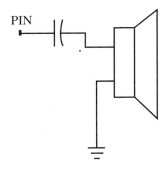

Figure 3-4: Circuitry required to generate sounds using the SOUND command.

What SOUND does is pulse the designated pin high and low at an audible frequency. The pulsing will continue for a length of time specified by the *duration* value. The values do not specifically tie into musical note values. The sounds produced fall into two categories, tones and white noise. Tones are values 1 to 127 and the higher frequency white noise values are 128 to 255. Value 0 is for silence. It can be used to produce a pause between notes or white noise.

Duration is a value of 0 to 255 measured in milliseconds. Additional notes and duration values can be include in a single command. With the right combination, a short melody can be produced. Using just a single note and duration makes it easy to produce feedback if a button is pressed. Here's a short program example:

```
SOUND 0, [100, 10, 50, 20, 100, 10, 50, 20]    'Make a cycling
                                                'sound that
                                                ''alternates
                                                'between note
                                                '100 and note
                                                '50 on PORTB
                                                'PIN 0. Each
                                                'note has a
                                                'different
                                                'duration.
```

STOP

This command just puts the program in an endless loop and does not put the PIC in low-power mode. This command is not that useful.

SWAP *var, var*

This command is used to swap the contents of two variables. The *var* describes the variables to use. If $B0 = 7$ and $B1 = 1$, then after using this command on those variables $B0 = 1$ and $B1 = 7$.

Example:

```
Swap B0, B1
```

TOGGLE *pin*

This command reverses the state of the port pin in the data register. If a port pin was high, it is changed to a low. If it was low, then it's changed to high. If the port pin was an input prior to this command, the port pin is made an output and then the state of that port pin in the data register is reversed. For example:

```
TOGGLE 2            'Change state of PORTB bit 2. (PIC pin 8 on
                    '16F84)
```

WHILE *condition*
statements ...
WEND

This is a grouped set that lets a group of PicBasic commands run while the *condition* value is true. The *condition* is any comparison statement such as <, >, =, <=, etc. This can be very handy for performing a function while a port is high or low or while a variable is below a value that is changing by an interrupt routine.

Example:

```
LED     var     portb.0

WHILE portb.1 = 1           'While Port B pin 1 is high
        LED = 1             'light LED on Port B pin 0
WEND
```

WRITE *address, value*

This command will only work on PIC devices with internal EEPROM, like the 16F84. The *address* variable is the location to write to. The value in the *var* variable will be stored at the specified location.

Here's an example:

```
loop:
        for b1 = 0 to 9             ' The first 10
                                    ' locations in EEPROM
        write b1, 0        ' are set to 0.
        next
```

XIN *datapin, zeropin, {timeout, label,}[var, var, ...]*
XOUT *datapin, zeropin, [housecode\keycode{\repeat}, housecode\keycode{\repeat}, ...]*

These commands are used to communicate with X-10 format modules available from various sources such as Radio Shack. These modules allow electronics to communicate a command signal over house wiring. The intention is to allow development of a "smart house." These commands are used to make the PIC the central control panel for various X-10 modules throughout a house or allow the PIC to be the front end to a household function.

The *datapin* and *zeropin* are the PIC pins used for the communication. The XOUT command sends data as a code, set up by the X-10 format. X-10 codes are usually predefined for the modules you are using and you will need to study the documentation for those X-10 modules. The XIN command allows you to make your own X-10 module and you can define what each code will do. It's similar to any other electronic communication except the communication is using the house wire.

I suggest you read the PBPro manual for these commands and also understand the X-10 format before experimenting with these. They work with the high-voltage house wiring which is very untypical for PIC projects.

Summary

This concludes our discussion about PBPro, but by no means is it a complete summary. My intent was to complement the PBPro manual. PBPro is so powerful and has so many features it could be a book by itself. My intention was to give you a good reference summary and also offer an alternate explanation of the commands.

I suggest you look at the "readme" file included on your version of PBPro. Any new commands or updates to PBPro since this book was written will be in that file.

Now we can start talking about the guts of a PIC and the fun stuff of actually using PBC and PBPro to program PICs and make those microcontrollers do something. So let's move on!

CHAPTER 4

Inside the PIC Microcontroller

As mentioned in the previous chapters, the PicBasic Compiler (PBC) and PicBasic Pro (PBPro) work with the 14-bit core family of PIC microcontrollers. Although PBPro can work with the 16-bit core devices, I'll be focusing on the 14-bit ones here. (Most of the PIC features are common between them anyway.)

In this chapter, I'll detail some of the most common features within these devices. If you are anxious to start using PicBasic, you can skip this chapter and read it after doing the projects. While the information in this chapter is really helpful to understand the PIC's features, the PBC and PBPro compilers make programming so easy you can skip this and still successfully write PIC software.

In projects where you need some of the detail in this chapter, I'll explain just the bare essentials to get you going on the project. If you really want to know all about PICs before programming, or if you are using the book to teach a PIC class, then this chapter should be covered before moving on to the projects.

Fundamentals

The PIC family has many cousins with various unique features. Despite this, Microchip designed the PICs to share many common attributes such as memory layout and packaging layout. For example, the pin-out for the 18-pin package is the same whether it's a device with standard I/O, or analog comparators, or even analog-to-digital converter ports. They also share the same set of core assembly language commands and memory locations for special function registers. What this means is that upgrading from one version to another with more features requires very little code "tear up" or hardware changes. In fact, some upgrades require no changes at all.

Because of the common set of core PIC attributes, I can cover the most common features and still have covered a lot of ground. You can then refer to the PIC data sheet for any specific details.

Program Memory

Program memory is the space where your PicBasic program resides. When you read a data sheet and see the PIC is 0.5k, or 1k, or 2k, they are referring to the program memory space. Those sizes—0.5k, 1k, 2k , 4k, and 8k—are the program memory sizes for the majority of the PIC devices. This may seem incredibly small, but for the functions a PIC is designed to do, it's not.

A PicBasic program can have several hundred commands and still fit in a 1k device. Because each PicBasic command is so powerful, a 1k program could control a motor with feedback and direction control, or monitor a burglar alarm system while displaying information on a liquid crystal display (LCD). Adding a serial port to send the status of the system sensors to a PC is as easy as a single SEROUT command.

It's possible to fit so much in because you are creating what is known as *embedded software*. There is no operating system like DOS, and no printer drivers or graphic drivers to include. The embedded software just controls the switching of the I/O ports to control the surrounding electronics that are connected to the PIC.

Program memory in a PIC is actually 14 bits wide even though PICs are considered 8-bit microcontrollers. The 8-bit terminology comes from the data memory buss, which is 8 bits wide. It's common to refer to the 14-bit-wide memory as bytes even though a byte by definition is 8 bits wide. Microchip calls the 14-bit-wide addresses "words" even though a word is 16 bits long. This gets very confusing when you try to compare the size of a PIC to another 8-bit micro. PBC and PBPro call the 14-bit-wide memory "words." When your program compiles successfully, PBC and PBPro will display how many words of program memory were used.

The 14-bit PIC program memory is sectioned off starting at location 0000h, or zero hex.

It really helps to understand binary and hex number systems to understand how to use a microcontroller and I'll assume you do. If you don't, you can still use PicBasic because it defaults to decimal mode, and that helps a lot.

On a 0.5k PIC, the memory goes from 0000h to 01ffh (0 to 511 decimal, 14-bit wide words). A 1k part is 0000h to 03ffh, and a 2k part is 0000h to 07ffh. Some PICs do offer more than 2k, but PBC works with the first 2k only. You'll need PBPro to use the larger memory PICs.

The reason for the PBC limitation is because of the way the PIC expands the program memory beyond 2k, using a *banking* scheme. This means they share the address connections between each 2k block, and then have one or two bits select which bank of 2k to read from. A compiler has a tough time knowing exactly where the compiled code will end up in the PIC memory space, so it has to either limit the program to the first bank or do some elaborate code testing to handle the banking. PBPro handles that.

Reset Vector

All 14-bit PICs share some very important program memory locations. The first is the RESET VECTOR. It always occupies the first byte, or 0000h. When the PIC is first powered up, it has to know where to start executing the program. The PIC does that by executing the instruction at location 0000h first.

That first instruction is usually the beginning of your program, but it can sometimes be a directive to jump to another part of your program. The PicBasic compilers take care of all that so you don't need to worry about this. It's just good information to know and can be used when you get to the advanced stages of PicBasic programming.

The PIC can be reset from various sources, including the MCLR pin. Depending on the type of reset, internal registers may be affected. That is why it is best to initialize all I/O registers and special function registers at the top of your program. Consult the PIC data sheet about resets for more detailed information on this subject.

Data Memory

Data memory is where all your variables are stored. This is RAM (*Random Access Memory*), which means when the PIC is disconnected from power all the data memory is lost. The data memory is 8 bits wide, which is why the PICs are considered 8-bit microcontrollers. The PIC data memory is also banked just like the program memory. The first section of data memory is reserved for special function registers. These registers are the key locations that control most of functions within the PIC.

Some of the registers are located in Bank 0 while some are located in Bank 1. Some are even repeated so they are available in both banks. To select which bank to control, a bit is set or cleared in the STATUS register. Because you need access to the STATUS register from both banks, it is located in both banks.

PicBasic makes it so easy for the beginner that in most cases you do not need to manipulate the STATUS register. The PicBasic commands take care of it for you. There may be times when you are doing something unique, and that will be the time you will manipulate the special function registers. You can easily do that with the PEEK and POKE commands in PBC or directly write to the register in PBPro.

STATUS Register

The STATUS register is located in data memory at location 03h. Most of the STATUS functions are useful if you're programming in assembly; they control which bank of memory you are working on, tell you when the PIC is fully powered up after a reset, and even indicate what the results were of a recent math function. I don't believe you will be using this register much, but the information is here for reference. The register looks like the following:

STATUS REGISTER

IRP	RP1	RP0	$\overline{TO}$	$\overline{PD}$	Z	DC	C

IRP—Register Bank Select bit (used for indirect addressing)

    0: Bank 0,1 (00h - FFh)

    1: Bank 2,3 (100h - 1FFh)

RP1:RP0—Register Bank Select bits (used for direct addressing)

    00: Bank 0 (00h - 7Fh)

    01: Bank 1 (80h - FFh)

    10: Bank 2 (100h - 17Fh)

    11: Bank 3 (180h - 1FFh)

$\overline{TO}$*—Time-out bit*

    1: After power-up timer

    0: After watchdog timer time-out

$\overline{PD}$*—Power-down bit*

    1: After powering up

    0: After entering sleep mode

Z—Zero bit

    1: The result of the last assembly command arithmetic instruction was zero.

    0: The result of the last assembly command arithmetic instruction was not zero.

DC—Digit Carry bit

    1: The lower order nibble overflowed a bit to the higher order nibble.

    0: No lower order nibble to higher order nibble occurred.

C—Carry bit

    1: The last assembly command overflowed the most significant 8th bit.

    0: The last assembly command did not overflow the most significant 8th bit.

In most PicBasic programs, the most you can do is change the RP1 and RP0 bits. The rest are only useful if you're inserting assembly language in your PicBasic program or writing in assembly.

I/O Registers

This is the set of registers that you will use most. Every I/O port has two registers that control its operation, a *data register* and a *direction register*.

The data register is named after the port it controls. Port B, for instance, has a register name of PORTB. Within that register each of the eight bits, or pins, are controlled. This register determines if the pins should be a high or a low when the port pins are set up as outputs. If the port pins are set up as inputs, then this register is where the level of the port pins is stored when the port is read.

The data direction register is called the TRIS register, meaning *Tri-State* register. "Tri" means three, and the three states are High, Low and High Impedance. Each I/O port has its own TRIS register named after the port letter: TRISA is for Port A, TRISB for Port B, etc.

Each pin of the port can be set up as an input or an output by controlling a bit within the TRIS register. A "1" makes the pin an input and a "0" makes the pin an output. You can arrange these ports in any combination of input and output. You can change the direction at any point in your program. With proper care, you can read a port as an input and then drive something from the same port as an output.

PBPro allows you to manipulate these registers directly, but PBC requires you to use the PEEK and POKE commands except for Port B. As mentioned earlier, PBC commands automatically act on Port B handling the TRIS register for you. (One of the projects I demonstrate in the following chapters is how to control the I/O pins.)

Within the PIC banked register memory, the Data registers are on Bank0 and the TRIS registers are on Bank1. Here is the list of port memory locations.

Port A: $05

TrisA: $85

Port B: $06

TrisB: $86

Port C: $07

TrisC: $87

Port D: $08

TrisD: $88

Port E: $09

TrisE: $89

Ports D and E are only on the larger PICs and can sometimes move a few other PIC registers around within the memory map. That is why the data sheet for the part you are using is important to have. It will have all the memory locations displayed for you.

A/D Registers

Analog-to-digital registers are used specifically to control the A/D ports. On most A/D equipped PICs, the A/D ports will be included on Port A only. Some of the larger PICs also add more A/D ports by using Port E.

The A/D structure in the PIC uses three registers for access and control: the "A/D control register 0" (ADCON0), the "A/D control register 1" (ADCON1), and the "A/D result register" (ADRES). The ADCON0 register is really more of a control register while ADCON1 is a setup register.

ADCON0

ADCS1	ADCS0	CHS2	CHS1	CHS0	GO/DONE	Not Used	ADON

ADCS1-0—A/D Conversion Clock Select

These two bits allow you to pick from four different clock sources. The clock signal is used in the sample and hold A/D circuitry inside the PIC. The best choice is the internal RC oscillator since it runs independent of the external crystal/ resonator.

The other choices are for more precise measurements and require a lot of specific calculations—more calculations than the average PicBasic user will want to deal with.

The bit selections are:

00: External Oscillator / 2

01: External Oscillator / 8

10: External Oscillator / 32

11: Internal RC Oscillator

CHS2-0—Analog Channel Select

These bits choose which A/D port you want to read within your program. You will have to select this at the beginning of your PBC A/D routine. PBPro automatically selects this when you use the ADCIN command.

CHS2-0 select as follows:

000: Channel 0 (A0 pin)

001: Channel 1 (A1 pin)

010: Channel 2 (A2 pin)

011: Channel 3 (A3 pin)

100: Channel 4 (A5 pin)

101: Channel 5 (E0 pin)

110: Channel 6 (E1 pin)

111: Channel 7 (E2 pin)

Notice how pin A4 is skipped. That is because A4 is used for the timer TMR0 input and also is an open source output. Any digital I/O on A4 requires an external pull-up resistor.

GO/DONE—A/D Conversion Status bit

This bit is really a control bit and an indicator flag. It is used to monitor when the A/D conversion is complete. It allows your program to check A/D status. When it is set to a "1", the A/D conversion process starts. This bit is automatically cleared when the conversion is complete.

Not Used—This bit is not used for anything and can be ignored.

ADON—A/D On bit

This bit turns the A/D circuitry on or off. Setting this bit to a "1" will enable the A/D converter at the channel selected in bits CHS2-0. Setting this bit to a "0" shuts down the A/D circuitry so it doesn't draw any current.

This bit has to be set before the GO bit is set. In fact, the PIC requires your program to delay one sample time period between turning the A/D converter on (ADON = 1) and starting the conversion (GO = 1). That time has to be calculated to be exact but it's usually less than 50 microseconds. PBPro takes care of this in the ADCIN command.

ADCON1

ADFM	Not Used	Not Used	Not Used	PCFG3	PCFG2	PCFG1	PCFG0

The ADCON1 register is where you set Port A to be digital or analog input mode. Remember, the TRISA register just makes the port an input or output. ADCON1 takes it the next step and selects what kind of input. This is only used on Port A and Port E because they share standard I/O circuitry with the A/D converter circuitry.

The ADCON1 bits are as follows:

ADFM—A/D Result Format

(This bit is only used on the larger PICs with 10-bit A/D converters. It selects how the result is stored in a 16-bit space.)

0: Left Justified

1: Right Justified

PCFG3-0—A/D Port Configuration Control

These bits set which A/D port pins are non-A/D digital and which are analog A/D converter pins. It also selects the voltage reference used by the A/D converter.

The list below is the shorter version used for 18-pin PICs with 8-bit A/D; check the data sheet for the larger PICs with more A/D channels.

PCFG	A0	A1	A2	A3	Vref
00	A	A	A	A	Vdd
01	A	A	A	Vref	A3
10	A	A	D	D	Vdd
11	D	D	D	D	----

ADRES

This register is where the result of the A/D conversion is stored. On larger 10-bit A/D PICs, this register is doubled to form the ADRESH and ADRESL registers. ADRESH is the high byte and ADREL is the low byte. If you are just using an 8-bit PIC, then all you need is ADRES.

ADRES and ADRESH are at the same memory location. It allows you to use similar code from an 8-bit A/D PIC for occasions when you want 8-bit results on a 10-bit A/D converter. One of the projects in the later chapters will demonstrate how to use the A/D converter in PBC and PBPro.

Peripheral Interrupt Vector

One of the features the 14-bit core PICs offer over the 12-bit core PICs is *interrupts*. An interrupt is an internal hardware circuit that, if enabled, will interrupt your program and jump to another program. Interrupts can come from various sources, depending on what the PIC offers. All interrupts can be disabled, or only selected interrupts can be disabled by clearing a control bit. This is known as *masking* an interrupt.

The PERIPHERAL INTERRUPT VECTOR is the location in the program memory where the PIC jumps to if an interrupt occurs. The memory location is 0004h. In this location you put a `goto` statement that redirects the program to your interrupt routine.

When that interrupt routine is complete, a `RETFIE` (return from interrupt enable) assembly command returns the program to the location it was at before the interrupt occurred. The actual selection of interrupts depends on the PIC device that you are using. There are timer overflow interrupts, change in state interrupts for I/O, external signal interrupts for I/O, interrupts for completion of a function like serial communication, and others. To really understand all the interrupt functions, you need to study the data sheet for the part you are using.

PBC does not support interrupts in Basic. To use interrupts with PBC, you need to insert some assembly code and modify one of the include files the PicBasic compiler uses to convert your PicBasic program to a hex file. This is really an advanced function and beyond the scope of this book.

PBPro does offer interrupts in Basic and can also work with interrupts in assembly. The PBPro manual has a section in the back that discusses this. The interrupts in Basic feature doesn't use all the capabilities of the PIC's internal interrupt structure, so it is a bit slower in responding than an assembly program would be. In most cases though, the PBPro "On Interrupt" command will perform all your interrupt requirements.

There are two internal registers you need to understand to use the internal interrupts: the OPTION register and the INTCON register.

OPTION Register

This register is used to set up various internal features of the PIC. Within this register you can control the timer and Watch Dog timer prescaler that is used to extend the time it takes for the internal TMR0 timer, or the WDT, to overflow.

It also has a bit for selecting the TMR0 timer clock source. The internal oscillator clock or an external source at the TOCKI pin (A4) on the PIC are the two options. This second choice is handy for using the TMR0 timer as a counter instead. The OPTION register also enables or disables weak pull-up resisters on PORTB. This is handy for connecting to a bank of switches or keypad.

The OPTION register is also used to set up the interrupt control on Port B bit0. Port B pin 0 is designated an external interrupt pin. That pin can be set up to suspend the main program operation and run the interrupt routine if the pin changes state. The Option register sets up the direction of that change of state. It can be a rising signal (low to high) or a falling signal (high to low). This is done by setting the INTEDG bit.

The register breaks down as follows:

OPTION REGISTER

RBPU	INTEDG	TOCS	TOSE	PSA	PS2	PS1	PS0

RBPU—PORTB Pull-up enable bit

    0 : PORTB pull-ups are disabled

    1 : PORTB pull-ups are enabled

INTEDG—Interrupt edge select bit

    0 : Interrupt on rising edge of PORTB RB0/INT pin

    1 : Interrupt on falling edge of PORTB RB0/INT pin

TOCS—TMR0 Clock Source Select bit

    1 : Increment on pulse at PORTA RA4/TOCKI pin

    0 : Increment on internal clock pulse

TOSE—TMR0 source edge select bit when driven by TOCKI pin

    1 : Increment on high to low transition

    0 : Increment on low to high transition

PSA—Prescaler assignment

　　1 : Prescaler assigned to WDT

　　0 : Prescaler assigned to TMR0

PS2—PS0 Prescaler rate select bits

Bit Values	TMR0 Rate	WDT Rate
000	1:2	1:1
001	1:4	1:2
010	1:8	1:4
011	1:16	1:8
100	1:32	1:16
101	1:64	1:32
110	1:128	1:64
111	1:256	1:128

　　Both PBC and PBPro require the Watch Dog Timer to be enabled for the PAUSE, SLEEP, and NAP commands. Modifying the prescaler will affect these. That's why it is good to understand the registers' functions so you don't set up one feature and disable another.

INTCON Register

This register has a lot going on in it. The PIC has several different interrupts and this register controls all of them.

GIE

All the interrupts can be enabled or disabled by setting or clearing the GIE bit in the Option register. After that, you can disable or enable specific interrupts by setting the proper enable bit.

EEIE

This bit enables the EEPROM write interrupt, which indicates when the PIC has completed writing to the EEPROM. EEIE enables and disables that function. This is the only interrupt that has the indicator flag somewhere else. The EEPROM interrupt flag is located in the EECON1 register.

TOIE and TOIF

Next you have the timer TMR0 overflow interrupt enable bit. The TMR0 is an internal register that can be incremented by the internal PIC clock or by an external signal on the TOCKI pin, which is shared with the RA4 pin. The TOIE bit enables and disables this interrupt.

The TOIF flag is set if the TMR0 register counts beyond 255. The TMR0 register will overflow back to 0 and keep counting to 255 again. The TOIF indicates the TMR0 register overflowed so you can make corrections for it in your code.

INTE and INTF

The Port B Pin 0 interrupt, which has some of its control located in the Option register, can be enabled or disabled with the INTE bit. The INTF flag will indicate if the interrupt occurred.

RBIE and RBIF

This is another Port B interrupt. It is affected by the change of state at Port B pins 4 thru 7. This interrupt is handy for monitoring a keypad. The PIC can go into sleep mode and when this interrupt occurs, the PIC will wake up and run the code you have written to respond to this interrupt. The RBIF flag indicates that the change occurred.

INTCON REGISTER

GIE	EEIE	TOIE	INTE	RBIE	TOIF	INTF	RBIF

GIE—Global Interrupt Enable bit

    0 : Disable all interrupts

    1 : Enable all un-masked interrupts

EEIE—EE Write Complet Interrupt Enable bit

    0 : Disable EE write complete interrupt

    1 : Enable EE write complete interrupt

TOIE—Timer Overflow Interrupt Enable Bit

    0 : Disable timer overflow interrupt

    1 : Enable timer overflow interrupt

INTE—Port B pin 0 Interrupt Enable Bit

    0 : Disable Port B pin 0 Interrupt

    1 : Enable Port B pin 0 Interrupt

RBIE—Port B Change Interrupt Enable bit

    0 : Disable Port B change interrupt

    1 : Enable Port B change interrupt

TOIF—Timer 0 (TMR0) Overflow Interrupt Flag

    0 : TMR0 did not overflow

    1 : TMR0 overflowed

INTF—Port B pin 0 Interrupt Flag

    0 : Port B pin 0 interrupt did not occur

    1 : Port B pin 0 interrupt occured

RBIF—Port B Change Interrupt Flag

0 : Port B change interrupt did not occur

1 : Port B change interrupt occured

As you can see, the PIC offers a lot of interrupt options. I don't want to dive too deeply into this because this book is more about PicBasic than PIC internals. You should understand how to use this register if you use any interrupts in PBPro. By reading or clearing the bits in the INTCON register, you can control your interrupt routine.

Summary

I've only covered the basics here. The larger PICs have more timers and more registers to control the added interrupts, serial ports, capture/compare function, serial peripheral controller, and other advanced features. You need to really study the data sheets if you want to understand all those PIC features.

My experience has taught me that actually using PIC microcontrollers is the best way to learn. After completing the projects in this book, you should have enough experience with PICs to dive into the advanced features on your own.

Now let's build PIC-based projects with PBC and PBPro.

CHAPTER 5

Simple PIC Projects

As mentioned in the previous chapters, the PicBasic compilers work with the 14-bit core family of PIC microcontrollers. In this chapter we will put PicBasic to work with one of the more popular PICs, the 16F876. It is a great device to play with because of its flash memory. "Flash" memory means it can be programmed over and over again without having to erase it under ultraviolet light. The 16F876 PIC also has in-circuit programming capability that allows you to reprogram the flash memory without removing it from the circuit.

For this book, I will be using a *bootloader* to program the PIC. A bootloader is a small section of code pre-programmed into the PIC that allows you to download your program via a serial port. There are various versions of a bootloader available, but I'm using MELOADER from microEngineering Labs since they are also the company that produces the PicBasic compilers.

You will see both a PBC and PBPro version for each project. The PBPro requires a special directive line to work with the bootloader. If you program with a standard programmer, then that line can be eliminated.

Let's get started.

Project #1—Flashing an LED

We'll begin with the easy project of flashing an LED on and off. While this seems simple, even experienced PIC developers often start with this just to make sure things are working properly. Sometimes, within a complex program, I will flash an LED on an unused I/O pin just to give visual feedback that the program is running.

In this example, we'll flash an LED connected to port B pin 0 (RB0). The PIC I/O can individually sink or source 25 milliamps (mA) of current. That is more than enough to drive an LED directly. The software will light the LED for one second, then shut it off for one second, and then loop around to do it again. While this program is simple, getting the LED to flash on and off verifies that you successfully wrote the program in PicBasic, compiled/assembled it, programmed the PIC, and correctly built the PIC circuit. That is a big first step, and why you'll find this first project very rewarding. Figure 5-1 shows the completed circuit board for this project.

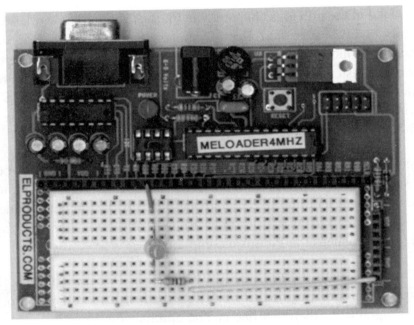

Figure 5-1:View of the completed circuit board for the LED flasher.

Figure 5-2 shows the schematic for this project; it will become the main building block for many of your PicBasic designs. It contains a 4-MHz resonator connected to the PIC OSC1 and OSC2 pins. It also has a 1k pull-up resistor from the MCLR pin to Vdd voltage of 5 volts. These are the only connections a PIC needs to run besides power and ground. That makes getting projects going much easier.

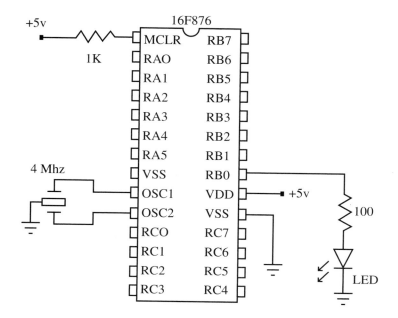

Figure 5-2: Schematic diagram of the LED flasher circuit.

PBC Code

The PBC code for this project is written to simply turn the B0 pin of Port B on and off at one second intervals. The first part of the program sets up the symbol LED to represent the Port B pin. This isn't necessary, but makes it easier to read then just putting a "0" everywhere the symbol LED is.

Next, Port B is set for which pins are inputs and which are outputs. Notice how PBC uses the DIRS directive and outputs are a "1" and inputs a "0". Finally, we enter the main code section. We use the `High` and `Low` commands to control Port B pin 0. The Pause command is used to create the one-second delay.

The main program runs in a continuous loop using the `GOTO` command to jump it back to the `main:` label.

```
' ---[ Title ]------------------------------------
'
' File...... proj1PBC.bas
' Format.... PicBasic
' Purpose... PIC16F876 flash LED
```

```
' Author.... Chuck Hellebuyck
' Started... June 16, 1999
' Updated...


' ---[ Program Description ]————————————————
' This is a simple program written to flash an LED by turning it
' on for one second and off for one second. The LED should be
' connected to portB pin 0 (16F876 pin 21) with the Cathode to
' ground and the anode at the PIC. A 100 ohm resistor is used in
' series with the LED to limit the current.

'---[ PBC Code ]————————————————

Symbol LED = 0              'Rename pin 0 of portb (PIC 16F84 pin 6)
                           'to LED

        DIRS = %00000001 'Setup port b as RB7-RB1 inputs, RB0 as
                           'output


main:                      'Label for beginning of main loop

High LED                   'Set pin 0 of portb high (5 volts) which turns
                         ' the LED on

        Pause 1000          'Pause 1000 milliseconds (1 second) with
                           'LED on

        Low LED             'Set pin 0 of portb low (0 volts) which
                           'turns the LED off

        Pause 1000          'Pause for 1 second with LED off

        goto main           'Jump to the main label and do it all
                           'again

        END                 'This line is not needed but its safe to
                           'put it here just in case the program gets
                           'lost.
```

PBPro Code

The code for the PBPro compiler is written to simply turn the B0 pin of Port B on and off at one-second intervals in a similar manner to the PBC version.

The first part of the program sets up the symbol LED to represent the Port B pin. This isn't necessary, but makes it easier to read than just putting a "0" everywhere the symbol LED is. Next, Port B is set for which pins are inputs and which are outputs. PBPro uses the TRIS directive and "1" is an input while "0" is an output. (This is a major difference between PBC and PBPro programs, as explained in the earlier chapters.) Finally, we enter the main code section. We use the High and Low commands to control Port B pin 0. The Pause command is used to create the one-second delay.

The main program runs in a continuous loop using the GOTO command to jump it back to the main: label.

```
'—-[ Title ]————————————————  —
'
' File...... proj1PRO.bas
' Format.... PicBasic Pro
' Purpose... PIC16F876 flash LED
' Author.... Chuck Hellebuyck
' Started... June 16, 2001
' Updated...


' —-[ Program Description ]————————————
' This is a simple program written to flash an LED by turning it
' on for one second and off for one second. The LED should be
' connected to port B pin 0 (16F876 pin 21) with the Cathode to
' ground and the anode at the PIC. A 100 ohm resistor is used in
' series with the LED to limit the current.

'—-[ PBPro Code ]————————————————

Define LOADER_USED 1      'Only required if bootloader used to
                          ' program PIC
```

```
symbol  LED = 0              'Rename pin 0 of portb (PIC 16F876 pin 21)
                            'to LED

       TRISB = %11111110    'Setup port b as RB7-RB1 inputs, RB0 as
                            'output

main:                        'Label for beginning of main loop

       High LED             'Set pin 0 of portb high (5 volts) which
                            'turns the LED on

       Pause 1000           'Pause 1000 milliseconds (1 second) with
                            'LED on

       Low LED              'Set pin 0 of portb low (0 volts) which
                            'turns the LED off

       Pause 1000           'Pause for 1 second with LED off

       goto main            'Jump to the main label and do it all
                            'again and again

       END                  'This line is not needed but its safe to
                            'put it here just in case the program gets
                            'lost.
```

Final Thoughts

There are several variations to this same program that will work; you may have a totally different format in mind. What I wanted to do with this was show the basics from which any program can be built. If you want to try other variations of this project, you could modify the values for "pause" to make the LED flash faster or slower. You could also add a second LED and alternate the on and off modes so the LEDs appear to flash back and forth. You could also just build the next project!

Project #2—Scrolling LEDs

This project expands on the previous project by lighting eight LEDs in a scrolling, "back and forth" motion. The entire Port B I/O is used to drive the eight LEDs. This is a good project that demonstrates how to control all eight LEDs with a single looping routine. The routine uses a `FOR-NEXT` loop to make it happen. The completed circuit is shown in Figure 5-3 and the schematic diagram is in Figure 5-4.

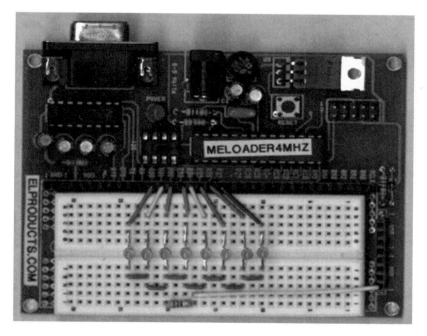

Figure 5-3: View of the completed circuit board for "scrolling LEDs" project.

Figure 5-4 shows the LEDs all sharing the same 100-ohm resistor. This works because only one LED is on at a time. Normally each LED should have its own resistor, and that would work here also without modifying the software. But I wanted to show another design approach that I thought was interesting. Notice that the same basic connections are used: OSC1, OSC2, MCLR, 5 volts, and ground.

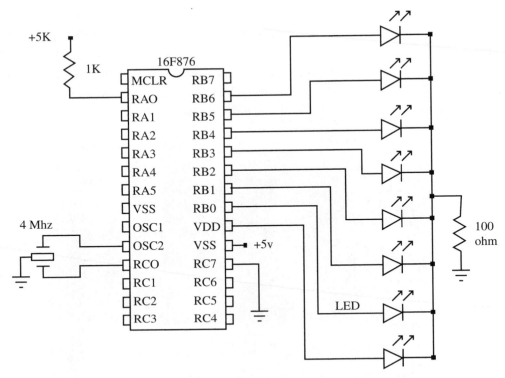

Figure 5-4: Schematic diagram of the "scrolling LEDs" circuit.

PBC Code

The PBC code for this project is very similar to that for the previous project, but expands on it by adding a FOR-NEXT loop to the code. It uses the same Port B to control the LEDs, but this time uses all eight ports of Port B to control eight LEDs.

In the PBC code, the symbol LED is established to represent the variable B0 rather than just a port pin. Through this variable LED, we will use the FOR-NEXT loop to increment through each Port B pin. Then we make all of Port B outputs with the DIRS directive and also add the PINS directive to establish all of Port B pins as low. This makes sure all the LEDs are off to start.

Next we enter the main loop of code. Within this main loop are two sub-loops created by separate FOR-NEXT loops. The first FOR-NEXT loop operates on the LED variable by increasing the value by one each time through the loop. This continues until LED is equal to 7, and then the second FOR-NEXT loop is entered.

The second FOR-NEXT loop actually initializes the LED variable to 7 and then decrements it by 1 on every loop. This continues until the LED variable is equal to 0. The LEDs light up in the opposite direction to the first FOR-NEXT loop. The program then has a GOTO statement to route the code back to the main: label to do it all again. The result is a scrolling light that moves back and forth.

This program shows how a small section of code within a FOR-NEXT loop can be used over and over to achieve different output results. The same effect could be achieved with a whole bunch of High and Low commands repeated for each LED. The effect would be the same, but the amount of program memory would be about five times larger!

```
' ---[ Title ]--------------------------------------------
'
' File...... proj2PBC.bas
' Format.... PicBasic
' Purpose... PIC16F876 scroll eight LEDs
' Author.... Chuck Hellebuyck
' Started... June 25, 1999
' Updated...


' ---[ Program Description ]------------------------------
' This program will scroll a string of LEDs in a back and forth
' motion. Each LED is turned on for '1 second and then turned off.
' The next LED in line is turned on immediately after the previous
' LED is turned off. This continues for all eight LEDs and then
' the direction is reversed. This creates a back and forth motion
' of light. All the LEDs are connected to port B which makes the
' code easier to implement.A single command can control all the
' LEDs at once by setting or clearing the bit associated with each
' LED.



'---[ PBC Code ]------------------------------------------

symbol      LED = B0            'Rename variable B0 as LED
```

```
       pins = %00000000    'Initiate all port B pins to low

       dirs = %11111111    'Setup port b as all outputs

main:                      'Label for beginning of main loop

' ********** Light LEDs in right direction ************

       for led = 0 to 7    'Loop through all LEDs

       high led            'Set each pin of portb high (5 volts)
                           ' which turns the LED on

       pause 1000          'Pause 1000 milliseconds (1 second) with
                           'LED on

       low led             'Set each pin of portb low (0 volts) which
                           ' turns the LED off

       next                'Continue until all 7 have been lit once

' ********** Light LEDs in left direction **************

       for led = 7 to 0 step -1      'Loop through all LEDs
                                     'backwards

       high led            'Set each pin of portb high (5 volts)
                           'which turns the LED on

       pause 1000          'Pause 1000 milliseconds (1 second) with
                           'LED on

       low led             'Set each pin of portb low (0 volts) which
                           'turns the LED off

       next                'Continue until all 7 have been lit once

' ********* Loop Back to Beginning **************

       goto main           'Jump to the main label and do it all
                           'again

       END                 'This line is not needed but its safe to
                           'put it here just in case the program gets
                           'lost.
```

142

PBPro Code

This PBPro code is also similar to that for the first project with an added FOR-NEXT loop. It uses the same Port B to control the LEDs, but this time uses all eight ports of Port B to control eight LEDs.

The B0 is not a predefined byte variable in PBPro as it is in PBC. Instead, we create a byte variable called "LED" using the VAR directive. Through this variable LED, we will use the FOR-NEXT loop to increment through each Port B pin. Next, we make all of Port B outputs with the TRISB directive and also add the Port B directive to establish all of Port B pins as low. This makes sure all the LEDs are off to start. Both of these directives act on the PIC register of the same name. This is more efficient than the way PBC handles this function.

Then we enter the main loop of code. Within this main loop are two sub-loops created by separate FOR-NEXT loops. The first FOR-NEXT loop operates on the LED variable by increasing the value by one each time through the loop. This continues until LED is equal to 7, and then the second FOR-NEXT loop is entered.

The PBPro format allows us to work directly on the Port B register to change individual bits. "PortB.0" is the portname.number format I mentioned earlier, and the line "PortB.0" represents pin 0 of Port B. We can add an equal sign after it and make it equal to 1 or 0. In this program, we take that a step further and add the "LED" variable to the end of it, "PortB.0[LED]". What that does is shift the bit to operate on from 0 to the LED value. For example, if LED equaled 5 then the "PortB.0[LED] = 1 would make the fifth bit of Port B high and turn on that LED.

The second FOR-NEXT loop actually initializes the LED variable to 7 and then decrements it by 1 on every loop. This continues until the LED variable is equal to 0. The PortB.0[LED] line then lights up the LEDs in the opposite direction from the first FOR-NEXT loop. The program then has a GOTO statement to route the code back to the main: label to do it all again.

This program shows how a small section of code within a FOR-NEXT loop can be used over and over to achieve different output results. The same effect could be accomplished with several PortB.0 = 1 and PortB.1 = 1, etc. commands repeated for each LED. As was the case with PBC code, the effect would be the same but the amount of program memory it would take would be about five times larger.

```
' —-[ Title ]————————————————————
'
' File...... proj2PRO.bas
' Format.... PicBasic Pro
' Purpose... PIC16F876 scroll eight LEDs
' Author.... Chuck Hellebuyck
' Started... June 25, 1999
' Updated...

' —-[ Program Description ]————————————
' This program will scroll a string of LEDs in a back and forth
' motion. Each LED is turned on for 1 second and then turned off.
' The next LED in line is turned on immediately after the previous
' LED is turned off. This continues for all eight LEDs and then
' the direction is reversed. This creates a back and forth motion
' of light. All the LEDs are connected to port B which makes the
' code easier to implement. A single command can control all the
' LEDs at once by setting or clearing the bit associated with each
' LED.

'—-[ PBPro Code ]————————————————


Define LOADER_USED 1     'Only required if bootloader used to
                         'program
' PIC


    LED var Byte         'LED variable setup as byte

      PortB = %00000000  'Initiate all port B pins to low

      Trisb = %00000000  'Setup port b as all outputs


main:                    'Label for beginning of main loop

' ********** Light LEDs in right direction ***************

      for led = 0 to 7   'Loop through all LEDs
```

144

```
portB.0[LED] = 1    'Set each pin of portb high (5 volts)
                    'which  turns the LED on

pause 1000          'Pause 1000 milliseconds (1 second) with
                    'LED on

portb.0[LED] = 0    'Set each pin of portb low (0 volts) which
                    ' turns the LED off
next                'Continue until all 7 have been lit once

' ********** Light LEDs in left direction ******************

for led = 7 to 0 step -1      'Loop through all LEDs
                              'backwards

portb.0[led] = 1    'Set pin 0 of portb high (5 volts) which
                    'turns the LED on

pause 1000          'Pause 1000 milliseconds (1 second) with
                    'LED on

portb.0[led] = 0    'Set pin 0 of portb low (0 volts) which
                    'turns the LED off

next                'Continue until all 7 have been lit once

goto main           'Jump to the main label and do it all
                    'again and again

END                 'This line is not needed but its safe to
                    'put it here just in case the program gets
                    'lost.
```

Final Thoughts

You can modify the timing to make the LEDs scroll faster or slower by changing the values for the "pause" command. With some trial and error, you could even get it to operate like the front of that car on the *Night Rider* TV show!

Project #3—Driving a 7-Segment LED Display

In this project, we drive a 7-segment display like those used in digital clocks and in old calculators. One of the biggest thrills for me, in my early days of fooling with electronics, was when I drove a 7-segment display to indicate 0 to 9 using four discrete integrated circuits. That was a long time ago! This project reduces those four chips to a single PIC.

Driving a 7-segment LED display is really the same as driving seven separate LEDs. Each segment of a 7-segment LED is an individual LED, but they have their cathodes (common cathode) or anodes (common anode) tied together at one pin. We'll use Port B of the PIC 16F876 to drive each segment individually of a "common cathode" 7-segment LED. The program will count from 9 to 0, and then light an LED on the eighth port of Port B that is not used by the 7-segment display. This program demonstrates the use of the Lookup command to set the proper Port B pins high to form the numbers 0 through 9. Figure 5-5 shows the completed circuit board for this project and Figure 5-6 gives the schematic diagram.

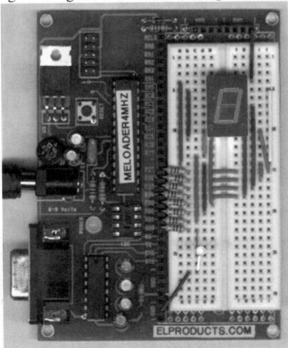

Figure 5-5:View of the completed 7-segment LED display driving circuit.

The circuitry uses a common cathode 7-segment display. The display pin numbers are not designated because I could not guarantee that the pin-out would match your LED display. You must check the pin-out data sheet for the display you use. The resistors for all the LED segments and the stand-alone LED are 100 ohms. The rest of the circuit is the standard PIC connections that were used in the previous projects.

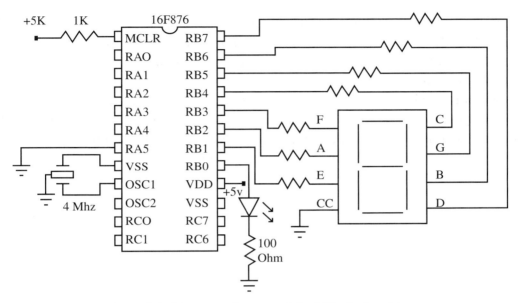

Figure 5-6: Schematic diagram of the LED display driver.

PBC Code

The PBC version of the project code starts as usual with the variables defined. A general purpose "x" variable is used as the count variable. A second variable "numb1" is used to store the segment arrangements that form the individual numbers.

Next the I/O is set up and all outputs reset to off. Because PBC defaults to working with Port B, the DIRS directive is used. The program then begins the main loop at label loop:. This loop is really formed by the FOR-NEXT commands. Within the FOR-NEXT loop, the variable "x" is decremented from its initial value of 9 to the lowest value of 0. Within that loop is also a command we haven't used in previous projects, the GOSUB command.

The GOSUB command forces the PIC to put the main loop on hold and run a second loop of commands. Those commands are at label convrt. The convrt loop contains another command we have not used, LOOKUP. This convrt loop takes the value of variable "x" and, through the LOOKUP command, stores the proper Port B I/O set-up in variable "numb1". The LOOKUP command constants, contained in the parentheses, are the Port B I/O arrangements required to display the number contained in variable "x" on the LED display.

The convrt loop returns back to the main loop using the RETURN command. It returns the PIC back to the command that follows the GOSUB command. The command line after the GOSUB command is where Port B actual drives the LED display with the value in "Numb1". It's done with the simple PINS directive. All this continues until variable "x" equals zero. Then the program leaves the FOR-NEXT loop and moves to the light label. Within that section, the program turns on the stand-alone LED in the same way Project #1 did earlier. After that, the program loops up to loop again and starts the whole countdown over.

```
' —-[ Title ]——————————————————
'
' File...... proj3PBC.bas
' Format.... PicBasic
' Purpose... PIC16F876 drives 7-segment LED
' Author.... Chuck Hellebuyck
' Started...  August 1,1999
' Updated...


' —-[ Program Description ]———————————
'This program drives a common cathode 7-segment LED display to
'countdown from 9 to 0 and then lights a separate LED for one
'second to signify the end of the countdown. The program then
'loops around and counts down again.

'—-[ PBC Code ]——————————————————


symbol      x = b0               ' Establish variable X

symbol      numb1 = b1           ' Establish variable numb1

init:
```

```
pins = %00000000          'Set all port B pins low
dirs = %11111111          'Set all port B as outputs

loop:
        for x = 9 to 0 step -1  'Countdown from 9 to 0

        gosub convrt            'Go to conversion routine

        pins = numb1            'Set proper I/O pins per convert
                                'routine

        pause 1000              'Keep display the same for 1 second

        next                    'Next number in countdown

light:
        high 0                  'Countdown reached 0 light LED

        pause 1000              'Keep LED lit for 1 second

        low 0                   'LED off

        goto loop               'Do it all again

'* Convert decimal number to proper segment alignment for LED
'display *

convrt:

lookup x,($DE,$50,$E6,$F4,$78,$BC,$BE,$54,$FE,$FC),numb1  'Match
                                          'segments to value of  x

return              ' Return to the line after the gosub

        end                      'Add this in case program gets lost.
```

PBPro Code

The PBPro code starts as usual with the variables defined. A general-purpose "x" variable is used as the count variable. A second variable, "numb1," is used to store the arrangement of segments that form the individual numbers.

Next the I/O is set up and all outputs reset to off at the `init` label. The program then begins the main loop at label `loop`. This loop is really formed by the FOR-NEXT commands. Within the FOR-NEXT loop is where the variable "x" is decremented from its initial value of 9 to the lowest value of 0. As with the PBC code, within that loop is also a command we haven't used in previous projects, the GOSUB command.

The GOSUB command forces the PIC to put the main `loop` on hold and runs a second loop of commands. Those commands are at label `convrt`. The `convrt` loop contains another command we have not used, `Lookup`. This `convrt` loop takes the value of variable "x" and, through the `Lookup` command, stores the proper Port B I/O set-up in variable "numb1". The `Lookup` command constants, contained in the brackets, are the Port B I/O arrangements required to display the number contained in variable "x" on the LED display.

Notice how the LOOKUP command in PBPro uses brackets around the selection list, while PBC earlier used parentheses. This minor code difference will produce an error if you try to convert a PBC program to PBPro.

The `convrt` loop returns back to the main `loop` using the RETURN command. It returns the PIC back to the command that follows the GOSUB command. The command line after the GOSUB command is where Port B actual drives the LED display with the value in "Numb1". It's done by directly modifying the Port B register.

All this continues until variable "x" equals zero. Then the program leaves the FOR-NEXT loop and moves to the `light` label. Within that section, the program turns on the stand-alone LED in the same way Project #1 did earlier. After that, the program loops up to `loop` again and starts the whole countdown over.

```
' ---[ Title ]----------------------------------------
'
' File...... proj3PRO.bas
' Format.... PicBasic Pro
' Purpose... PIC16F876 drives 7-segment LED
' Author.... Chuck Hellebuyck
' Started... June 17, 2001
' Updated...
```

```
' —-[ Program Description ]————————————————————
'This program drives a common cathode 7-segment LED display to
'countdown from 9 to 0 and then lights a separate LED for one
'second to signify the end of the countdown. The program then
'loops around and counts down again.

Define LOADER_USED 1      'Only required if bootloader used to
                          'program PIC

x var byte           ' General purpose variable

numb1 var byte        ' variable to store the 7-segment I/O setup

init:

portb = %00000000         'Set all port B pins low

trisB = %00000000         'Set all port B as outputs

loop:
      for x = 9 to 0 step -1  'Countdown from 9 to 0

      gosub convrt                'Go to conversion routine

      portb = numb1               'Set proper I/O pins per convert
                                  'routine

      pause 1000                  'Keep display the same for 1 second

      next                        'Next number in countdown

light:
      high 0                      'Countdown reached 0 light LED

      pause 1000                  'Keep LED lit for 1 second

      end

      low 0                       'LED off

      goto loop                   'Do it all again
```

```
'* Convert decimal number to proper segment alignment for LED
'display *

convrt:
lookup x,[$DE,$50,$E6,$F4,$78,$BC,$BE,$54,$FE,$FC],numb1 'Match
                                        'segments to value of x

        return

        end                'Add this in case program gets lost.
```

Final Thoughts

You could modify the count direction to count up instead of counting down just by changing the FOR-NEXT loop values from "9 to 0" to "0 to 9 and remove the step 1." You could also replace the stand-alone LED with a low current buzzer that will beep when the count is finished.

CHAPTER 6

Moving On with the 16F876

Now we'll discuss additional interesting projects that use more of the PIC's resources. We will continue to work with the 16F876 flash PIC and will be accessing additional I/O directly using PBPro and indirectly using the PEEK and POKE commands in PBC. After that, we use some of the special I/O of the 16F876 that includes an analog-to-digital (A/D) converter shared with the Port A digital I/O pins. With PBC, the A/D ports can be accessed with a few fairly simple commands and PBPro allows A/D access with a single command. As a final project, we'll drive a servomotor. These are popular with hobbyists and are really quite easy to control with both PBC and PBPro.

Some of these projects will access sections of the PIC described in Chapter 4. If you skipped that chapter, you might want to read it before attempting these projects.

Project #4—Accessing Port A I/O

Most of the PBC commands work directly on the pins of Port B, but what if you want to use another port in the PIC to do something? The answer is the POKE and PEEK commands. Through these commands you can change the state of the I/O pin from high to low or low to high or make a pin an input and read it for a high level or low level.

Because we are controlling I/O indirectly through PEEK and POKE, you have to know something about the inner workings of a PIC I/O structure. Every I/O port has two registers associated with it; a direction register called the TRIS (TRISA for port A) register and a data register called by the port name (i.e., PORTA) register.

When the PIC is first powered up, all I/O is put into a high impedance input mode. To make a pin within the port an output requires you to clear the bit associated with it in the port's TRIS register.

Bit 0 in the TRISA register determines the direction for the RA0 pin of PORTA. If the TRISA bit is a "0," then the pin is set to an output. If the TRISA bit is set to a "1," then the pin is set to an input. Therefore, when the PIC is first started up all the TRIS pins are set to "1" meaning all inputs. You can change any single bit to an output or change all of them to outputs.

In this project we will set PORTA pin 0 to an input to read the state of a switch. We will set PORTA pins 1 and 2 to outputs to drive LEDs. If pin 0 is high (switch open), we will light the LED on pin 1. If pin 0 is low (switch closed), then we will light the LED on pin 2. Seems easy enough, right?

(Note: Port A direction bits are opposite the direction bits PBC uses in the DIRS control of Port B. That's because PBC tried to maintain compatibility with the Basic Stamp. PBC inverts the DIRS to the proper Port B TRISB settings.)

PBPro is much simpler to use than PBC because it can operate directly on Port A the same way it operates on Port B in previous chapter projects. PBPro can access the TRISA register directly with a one-line command (TRISA = %00001111). Although the PBPro command set includes the PEEK and POKE commands, the compiler manufacturer does not recommend using them.

Once you look at the PBPro code and the PBC code described for this project, you'll understand how PBPro improves on the PBC structure and thus makes programming PICs in Basic easier.

The completed circuit is shown in Figure 6-1 and its schematic is in Figure 6-2. As you can see, the circuit is fairly simple. It contains a standard 4-MHz resonator connected to the PIC OSC1 and OSC2 pins. It also has a 1k pull-up resistor from the MCLR pin to the Vdd voltage of 5 volts. In addition, there are the connections required for the I/O control of Port A.

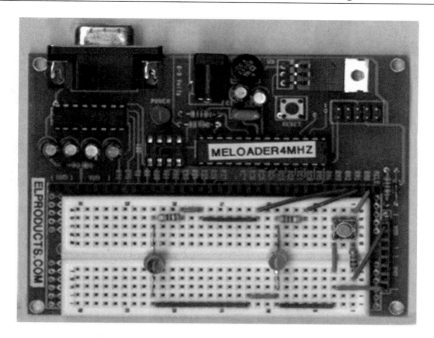

Figure 6-1: Completed circuit for the accessing port A I/O.

Notice that a connection to Port A doesn't look any different than the previous project's connections to Port B. The only difference between Port A and Port B in PBC is the way they are accessed in software. PBPro treats them the same.

The PIC has very powerful I/O circuitry that allows it to drive LEDs directly. A series resistor is required to limit the current. LED1 and LED2 are driven this way through 470-ohm resistors; changing them will change the LEDs' brightness. Just don't exceed 20 mA to prevent damage to the PIC.

The pushbutton switch is connected directly to Port A pin 0 (RA0). A 1k resistor "pulled-up" to 5 volts is also connected to it. The pushbutton switch is a normally open type that means the switch is open circuit when it's not being pressed. The 1k resistor sets the RA0 threshold to 5 v, or high, when the switch is not pressed.

When the switch is pressed, the signal at RA0 goes to ground indicating a low level. Make sure you use the correct type of switch when you build this or the circuit will not work properly with the software.

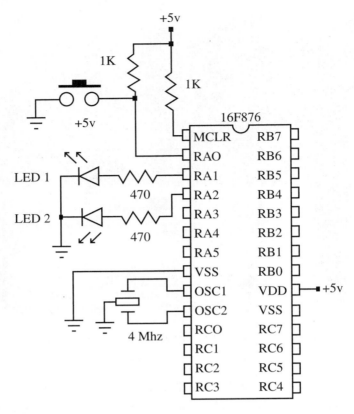

Figure 6-2: Schematic for circuit shown in Figure 6-1.

PBC Code

The circuit's PBC software, as listed below, first establishes the symbols Port A, TrisA, and ADCON1 by setting them equal to the memory location where they reside in the PIC. This makes understanding the PBC code much easier if you ever have to refer back to this code at a later date. After you use a PIC for a while, you will find you remember these location addresses but it's still much easier to read the port description.

If you're wondering where those memory location values came from, then look no further than the PIC data sheets or data book. Within those data sheets is a *memory map*. The memory map shows the numeric location of every register in the PIC.

The next set of instructions actually work on those locations using the POKE command. The digital direction of Port A needs to be set up by modifying the TrisA register. Notice how each port pin is set to a 0 or a 1 using the "%" binary directive. This makes it easy to tell which pin does what. The left-most digit is port pin 7 and the right-most is port pin 0.

The digital state of each pin is also established by modifying the Port A register in the same way the TrisA register was set up. Making a pin a 0 sets that pin low, a 1 makes it high.

In the line below, pin 1 of Port A is set high while the rest of the port is set low. This presets the LED connected to RA1 pin to the "on" or lit state.

```
poke PortA, %00000010          ' Set PortA RA1 high to turn
                               ' on LED1
```

Now I have to throw a little curve in here because I chose to use the PIC16F876. This PIC has an A/D shared with Port A. At power-up, the A/D has control over Port A. For that reason, the command to POKE a value of 6 into the ADCON1 register is necessary to set Port A up as a digital I/O port. Reading the data sheet about the A/D register will reveal this, but it's not instantly obvious.

The next section of code at label Main is the main loop. It first uses the PEEK command to read all the pins of Port A as a byte and store it in RAM byte prede-fined variable B0. Even though Port A pin RA1 is set to an output, it will read that pin as the state defined by the Port A register. RA2 will be read the same way. The rest of the pins of Port A are read as the voltage applied to those pins.

Since we only care about the state of RA0 where the switch is connected, the next line uses the & directive or AND directive to perform a logical AND on the byte that was read into RAM byte B0. By comparing it to %00000001, we are essentially erasing all the bits to 0, except for the last bit that is the state of the switch. If the switch is open, this bit will be high. If the switch is closed, this bit will be low. Therefore, if the switch is open B0 will equal one, and if the switch is closed B0 will equal zero.

The next line of code tests for that. By using the IF-THEN command we test for a zero value of B0. If B0 equals zero (switch closed), then we want the program to jump to label LED2. If B0 does not equal zero, then the program just jumps to the next command.

The next command repeats what was done at the top of the program and "pokes" LED1 on by setting that bit in the PORTA register. After the program completes that, the next command simply jumps the program back to the label Main to start the process all over again. On the other hand, if the value of B0 was indeed 0, then at label LED2 we use the same POKE command but this time to set Port A pin RA2 to a 1 and RA1 to a 0 which turns LED1 off and turns LED2 on to indicate the switch was pressed.

Immediately after that, the program jumps back to Main and tests the switch again. If the switch is not still being pressed, the program will turn off LED2 and turn LED1 back on with the PEEK command.

To keep the LED2 on, you have to keep your finger pressing the switch closed. As soon as you lift your finger off the switch, the LEDs should change back to LED1 on and LED2 off.

```
' ---[ Title ]----------------------------------------
'
' File...... proj4PBC.bas
' Format... PicBasic
' Purpose... Accessing PortA using Peek and Poke
' Author.... Chuck Hellebuyck
' Started... May 1, 2000
' Updated...


' ---[ Program Description ]------------------------
' This program demonstrates how to access PortA using Peek and
' Poke in PicBasic.


Symbol   PortA = 5                        ' PortA address
Symbol   TrisA = $85                      ' PortA data direction
                                          ' register
Symbol   ADCON1 = $9F           ' A/D control register (PortA
                                ' secondary control)
Init:
        poke TrisA, %00000001             ' Set PortA RA4-RA1 to
                                          ' output, RA0 to input
```

```
        poke PortA, %00000010            ' Set PortA RA1 high to
                                         ' turn on LED1

        poke ADCON1 = 6                  ' Set PortA to digital
                                         ' I/O

Main:
        ' *** Test the switch state ***
        peek PortA, B0            ' Read all PortA states and
                                  ' store in B0
        B0 = B0 & %00000001        ' Clear all bits in B0 except
                                   ' bit 0
        if B0 = 0 then led2        ' If switch is closed then
                                   ' jump to
                                   ' the LED2 on routine

         poke PortA, %00000010     ' Turn LED1 on, LED 2 off

         goto Main                 ' Jump to the top of the main
                                   ' loop

LED2:
        '*** Turn LED2 on ***
        poke porta, %00000100 ' LED2 on and LED1 off

         goto Main                 ' Jump to the top of the main loop
```

PBPro Code

The PBPro software, as listed below, first goes to work on the special function registers TRISA, PORTA, and ADCON1. PBPro makes it easier on you to access these registers because they don't require you to know the register address in the PIC. PBPro already has that built in.

The work done on those registers is to preset them to defined values by directly equaling them to a value. The digital direction of Port A needs to be set up by modifying the TrisA register. Notice how each port pin is set to a 0 or a 1 using the % binary directive. This makes it easy to tell which pin does what. The left-most digit is port pin 7 and the right-most is port pin 0.

The digital state of each pin is also established by modifying the Port A register in the same way the TrisA register was set up. Making a pin a 0 sets that pin low, a

1 makes it high. In the code below, pin 1 of Port A is set high while the rest of the port is set low. This presets the LED connected to RA1 pin to the on or "lit" state.

Now I have to throw a little curve in here because I chose to use the PIC16F876. This version of the PIC has an A/D shared with Port A. At power-up, the A/D has control over Port A. For that reason, the command to preset a value into the ADCON1 register is necessary to set Port A up as a digital I/O port. Reading the data sheet about the A/D register will reveal this but it's not instantly obvious.

The next section of code at label Main is the main loop. The first thing we do is go to work on the pin connected to the switch, RA0. PBPro allows us to use the IF-THEN command to test that pin directly using the portname.pinnumber convention. If the state of RA0 is high or a "1," then the switch is open and we should proceed to the next command. The next two commands work directly on Port A register to set RA1 on and RA2 off, thus turning LED1 on and LED2 off. After that, the program loops back to test the switch again. If the switch is closed when we were at the IF-THEN command, then RA0 will be low or a "0" and the command then directs the program to jump to label LED2.

At LED2 we operate directly on the Port A register to set and clear the LED pins to their proper state. In this case we set LED1 off and LED2 on. After that, the program loops back to Main to start all over again testing the switch. To keep LED2 on, you have to keep your finger pressing the switch closed. As soon as you lift your finger off the switch, the LEDs should change back.

If you compare the PBC program to the PBPro program, it becomes clear that not having to PEEK and POKE makes the PBPro version much easier to use and understand.

```
' ---[ Title ]----------------------------------    --
'
' File...... proj4PRO.bas
' Format.... PicBasic Pro
' Purpose... Using Porta on PIC16F876
' Author.... Chuck Hellebuyck
' Started... June 1, 2000
' Updated...
```

```
' ----[ Program Description ]----------------------------
'This program demonstrates how to control PortA with PicBasic Pro.
'Peek and Poke commands are not required because PicBasic Pro has
'control over all registers including I/O registers.
'direct This program will do the same function as proj4PBC.bas but
'in Pro format. PortA RA1 and RA2 will drive LEDs. The RA0 port
'will be an input and read the state of a momentary push button
'switch to determine which LED to light. If switch is pressed LED2
'will light. If switch is not pressed then LED1 will light.

Define LOADER_USED 1      'Only required if bootloader used to
                          'program PIC

Init:
    adcon1 = 6                          ' Set all PortA to digital
                                        'I/O
    trisa = %00000001                   ' set PortA RA4-RA1 to
                                        'outputs, RA0 input
    porta = %00000010                   ' Set PortA RA1 high to turn
                                        'on LED1

Main:
        ' *** Test the switch state ***
        if portA.0 = 0 then led2        'If switch is pressed then
                                        'jump to LED2 routine

        PortA.1 = 1             ' Turn LED1 on
        portA.2 = 0             ' Turn LED2 off
        goto Main              ' Jump to the top of the main loop

LED2:
        '*** Turn LED2 on ***
        porta.2 = 1             ' LED2 on
     porta.1 = 0             ' LED1 off
        goto Main                  ' Jump to the top of the main
                                   ' loop
```

Final Thoughts

It is obviously more complicated to access Port A than Port B in PicBasic. But after you set up variables for the TRISA register and Port A register locations, the program becomes easier to understand. POKE and PEEK are very useful commands since they allow your PBC program to access any register on the PIC. As you get more familiar with PICs, you will be able to access internal timers, analog-to-digital conversion, internal registers, and many other features available on the various PICs.

The PBPro compiler once again makes programming the PIC a little shorter and a little easier. Although these programs were simple, it was interesting to see how you could modify their function just by pressing a switch. You can build on this basic arrangement to add more switches and functions to control more than just LEDs. We'll touch on some of those in later chapters. But for our next project, let's skip the step where we changed the ADCON1 register to digital I/O and use Port A as an A/D converter.

Project #5—Analog-to-Digital Conversion

This project uses one of the most useful features of the PIC16F876, the analog-to-digital (A/D) converter. Almost everything in the real world is not digital but instead analog. To control something in the real world, or to understand something in the PIC, we have to convert that real-world analog data into the digital form the PIC understands. That is done with an A/D converter. F or example, if you have to read a temperature or light levels, you will need both a sensor to convert the measurement into a variable voltage and an A/D converter to change the resulting voltage into a digital value.

In this example, our sensor will be a variable resistor called a *potentiometer* (POT). As we turn the POT's shaft, we want to read the variable resistance from that POT and light some LEDs to show how much we turned it. This could be compared to the volume adjustment you make on a stereo. As you turn the knob for volume, the sound from the stereo gets louder. That's because it is reading the resistance of the POT connected to the knob you turned to adjust the amplifier's output.

This project will use a POT connected to Port A pin RA2. We'll control five LEDs using Port B. The program will have thresholds of A/D values associated with each LED so, as we turn the POT, the LEDs will light in order just like a bar meter on a stereo.

Fortunately, we don't need to know too much about the operation of the A/D circuit; the software just assumes the circuit works, and it does. A/D circuits come in different forms but they all do the same thing—convert an analog voltage into a digital voltage. An A/D register's digital output will have a *resolution* to it. That means it can output an 8-bit digital value, 10-bit digital value, or larger if required. The PIC 16F876 has a 10-bit resolution A/D register, but can also operate as an 8-bit. We will use it as an 8-bit since it's a little easier to understand. Eight bits fit into one byte, and that's much easier to manipulate in code.

Figure 6-3 shows the completed circuit and the schematic is given in Figure 6-4. The LEDs are connected to Port B with 100-ohm series resistors. Note that the same basic connections are used—OSC1, OSC2, MCLR, 5 volts, and ground—that were used in the previous project.

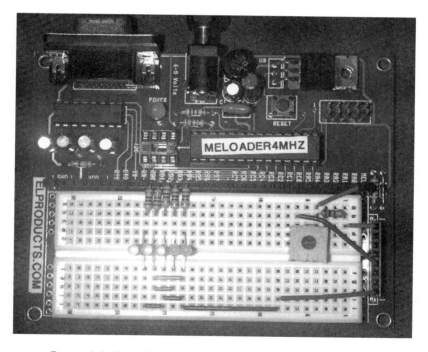

Figure 6-3: Completed circuit for the A/D conversion project.

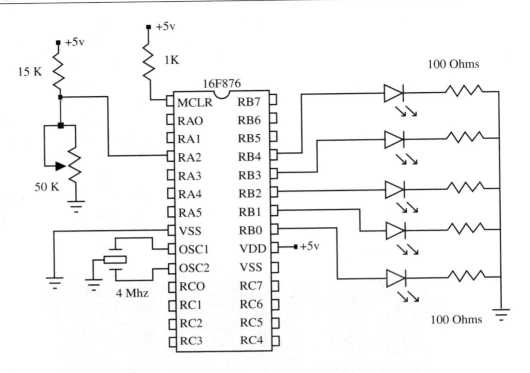

Figure 6-4: Schematic for the circuit shown in Figure 6-2.

Looking at Figure 6-4, notice we add the potentiometer to Port A RA2. The 15k pull-up resistor is needed to supply power to the POT. You can vary the values of the POT and pull-up resistor and still get similar results. By adjusting the POT, we are changing the voltage at RA2 through the resistor divider formed between the 15k resistor and POT.

The A/D port cannot handle voltages above 5 volts. Therefore if you need to measure larger voltages, you either have to step it down using resistors or build a voltage conversion circuit using an op amp IC. (But that's a subject for another book and another author!)

PBC Code

In the PBC program for this circuit, the first part of the PBC code establishes names for the register locations. ADCON0 and ADCON1 are special function registers for controlling the A/D register and ADRESH and ADRESL are where the result of the A/D conversion is stored. If you operate in 10-bit mode, both ADRESH and

ADRESL are used to hold the 10-bit result. If you use 8-bit resolution, then you only need one byte to store the result so that register is ADRESH. These register values are once again found in the PIC16F876 data sheet and I described them in Chapter 4.

After the A/D special function registers are established, the code then initializes the Port B pins at the `Init` label. Using the `PINS` directive and the `DIRS` directive, Port B is set as all outputs and all pins set to zero. Then the program enters the main code at label `Start`.

First step at `Start` is to set Port A to operate as an A/D register. The `POKE` command is used to adjust the TrisA register to all inputs. The `POKE` command is then used to set ADCON1 to hex value $02. This sets all Port A pins connected to the A/D converter to operate as A/D input pins rather than digital I/O pins. This adjustment of ADCON1 also clears the ADFM bit, described in Chapter Four, to set the A/D output to 8-bit mode. This will put the full result in the ADREH register as a byte.

Next we set the A/D converter operating mode. This step selects which A/D port to actually read. We `POKE` the ADCON0 register to control this. If we were reading more than one POT, we would have to do this step over again for that second POT connected to a separate A/D port.

We set ADCON0 to %11010001. Starting from the left, the two most significant bits select the RC internal oscillator as the clock for the A/D circuitry. (Unless you are into extremely accurate A/D measurements, this is the choice to use.) The next three bits are "010" and they select channel 2 or pin RA2 as the A/D port pin to read. Finally, the last bit is a "1" and it turns the A/D converter on. In fact, it starts the A/D conversion.

Since the A/D conversion is started, we have to check when it is completed. We do that at label `loop`. We test it by first using `PEEK` to copy the whole ADCON0 register into predefined RAM byte B0. Then in the next command line we use and `IF-THEN` statement to test the Go/Done bit by using the predefined "Bit2" name associated with RAM byte B0. That's the advantage of using B0 as the register because we have easy access to the bits. If that bit2 value is 1, the conversion is not complete so we loop back to `PEEK` ADCON0 again. If the bit2 value is 0, then we know the A/D conversion is complete and the result is stored in the ADRESH register. We then use the `PEEK` command to read ADRESH and store the result in predefined RAM byte B3.

Now that we have the A/D result from the POT, we need to determine how many LEDs to light up. We do that with a series of IF-THEN statements. We test the value four times to see if it is greater than a preset value. This program tests it for a value of 25, 75, 125, and 175. If the value of B3 is less than the predefined value, the next command uses the Pins directive to set the correct number of LEDs. If the value of B3 is larger than the predefined value, the IF-THEN command jumps the program to the next IF-THEN test of B3. If all the IF-THEN commands are smaller than the value of B3, the final step simply sets all the LEDs on because the value is greater than all the tested values.

After the LEDs are set, we pause 100 milliseconds to let the LEDs glow. Then the program loops back up to Loop to get another reading off the POT. The 100-millisecond delay can be eliminated or reduced if you want to read the POT a little faster.

To understand how the voltage at the POT is compared to the set values in the IF-THEN section of code, let's look at the math involved. The A/D converter defaults to the 5-volt Vref as the reference voltage used internally by the A/D converter. It takes the ratio of the voltage at RA2 and the Vref voltage of 5 volts and multiplies it by 255. The result of that calculation is then stored in the ADRESH register. For example, if the voltage at RA2 is 2.30 volts, then the result would be:

$(2.30 / 5) * 255 = 117.30.$

In our code we turn on the first three LEDs when the A/D result is less than 125.

```
tst3:
        if B0 > 125 then tst4   'If A/D value is between 75 and 125
        pins = %00000111               ' then light LED0 - LED2
        goto cont                 'continue with the program
```

Therefore the fourth LED will light when the voltage at RA2 increases above the 125 value, which is just about 2.45 volts.

To me this is a great example of how programming in PicBasic is so powerful. You are using simple PEEK and POKE commands to control a high-powered microcontroller the same way someone programming in assembly code would do. But it's so much easier to read and understand. You will see that the A/D converter is very accurate and not too difficult to set up. That's why in earlier chapters I didn't recommend the POT command because using a PIC with A/D gives more accurate and consistent results.

```
' ---[ Title ]--------------------------------  --
'
' File...... proj5pbc.BAS
' Purpose... POT -> 16F876 -> LEDs
' Author.... Chuck Hellebuyck
' Started... May 19, 2001
' Updated...



' ---[ Program Description ]-------------------  -
'
' This Program uses the 16F876 to read a potentiometer (POT) and
' drive LEDs in a bar graph mode as the POT is turned.
'
' RA2        pot connection
' RB4        LED4
' RB3        LED3
' RB2        LED2
' RB1        LED1
' RB0        LED0



' ---[ Revision History ]--------------------------
'
'


' ---[ Constants ]---------------------------------  --
'


' A/D Variables and symbols
'
Symbol  ADCON0 = $1F                          ' A/D Configuration
Register 0
Symbol  ADRESH = $1E                          ' A/D Result for 8-bit
mode
Symbol ADRESL = $9E
Symbol  ADCON1 = $9F                           ' A/D Configuration
Register 1
Symbol  TRISA  = $85                           ' PortA Direction
register
```

```
' ---[ Variables ]----------------------------------
'
' B0 and B3 are used but predefined in PBC therefore no symbol
' required

' ---[ Initialization ]-----------------------------
'
Init:
        pins = $0000                        ' all outputs off to
                                            ' start
        Dirs = %11111111                    ' All of Port B is
                                            ' outputs



' ---[ Main Code ]----------------------------------
'

'*********** A/D Read ***********************
'
' PEEK and POKE Commands
'
' Access 16F876 A/D using Peek and Poke

Start:
        poke TRISA, $FF                     'Set PortA to Inputs
        poke ADCON1, $02                    'Set PortA 0-5 to analog
                                            ' inputs, and also
                                            ' Sets result to left
                                            ' justified 8-bit mode
        poke ADCON0, %11010001              ' Set A/D to RC Osc, Channel
                                            ' 2, A/D converter On

loop:
        Peek ADCON0,B0
            Bit2 = 1
            Poke ADCON0,B0                  ' Set ADCON0-Bit2 high
                                            ' to start conversion

test:
        Pause 5
            Peek ADCON0,B0
            If Bit2 = 1 Then test           ' Wait for low on bit-2
                                            ' of ADCON0, conversion
                                            ' finished

        Peek ADRESH,B3                      ' Move HIGH byte of
                                            ' result B3 variable
```

```
'********** Drive LEDs ************************
LEDtst1:
        if B3 > 25 then tst2        'If A/D value is less than 25
        pins = %00000001            ' then light LED0 only
        goto cont                   'continue with the program
tst2:
        if B3 > 75 then tst3        'If A/D value is between 25
                                    ' and 75
        pins = %00000011            ' then light LED0 & LED1
        goto cont                   'continue with the program
tst3:
        if B3 > 125 then tst4       'If A/D value is between 75
                                    ' and 125
        pins = %00000111            ' then light LED0 - LED2
        goto cont                   'continue with the program
tst4:
        if B3 > 175 then tst5       'If A/D value is between 125
                                    ' and 175
        pins = %00001111            ' then light LED0 - LED3
        goto cont                   'continue with the program
tst5:
        pins = %00011111            'A/D value is greater than 175
                                    ' so
                                    ' light all the LEDs 0-4
cont:
        Pause 100                   'wait 1 second
        goto loop


        end
```

PBPro Code

And now for a real demonstration of the simplicity the PBPro compiler offers, we'll use the ADCIN command to perform the same function that took several steps in PBC. We'll read the same A/D port RA2 and light the same LEDs based on the same threshold values, but we'll do it in about half the code space PBC required.

We start off with the DEFINE statements required by PBPro. The same Loader_Used define statement is at the top because I am using a bootloader to program the PIC. Next are a series of DEFINE statements dedicated to the ADCIN command. These make it simple to set the output result to eight bits, the clock source to RC, and add a sample time that sets when we check the status of the A/D conversion.

Next, at the init label the program establishes a byte variable called "adval." This is where the A/D result will be stored. After that, we set Port B to be all outputs and initialize all LEDs to off by setting all the Port B pins to 0.

After that we enter the main code section. We first set Port A to all inputs by modifying the TrisA register. Then we work on the ADCON1 register to make all inputs of Port A work with the A/D register rather than as digital I/O. Then the ADCIN command is issued. Within this command we define which A/D port to read (2) and where to put the result (adval). The testing for completion of the A/D conversion is all done by the ADCIN command. After the conversion is complete, we can go right to work on the result and test it against the IF-THEN statement thresholds the same way the PBC version did.

At each step, we compare "adval" to the predefined values. We light the LEDs by working directly on the Port B register. We light the LEDs according to the value of "adval." If "adval" is less than a predefined value in the IF-THEN statement, then the next command is the Port B manipulation. If none of the values in the IF-THEN statements are larger than "adval," then the Port B register is changed to light all five LEDs. After this, the program pauses for 100 milliseconds and then jumps back to the Loop label to test the A/D register again.

```
'  —-[ Title ]————————————————————  --
'
'  File...... proj5pro.BAS
'  Purpose... POT -> 16F876 -> LEDs
'  Author.... Chuck Hellebuyck
'  Started... May 19, 2001
'  Updated...
```

```
' ---[ Program Description ]------------------------

'
' This Program uses the 16F876 to read a potentiometer (POT) and
' drive LEDs in a bar graph mode as the POT is turned.
'
' RA2          pot connection
' RB4          LED4
' RB3          LED3
' RB2          LED2
' RB1          LED1
' RB0          LED0



' ---[ Revision History ]------------------------
'
'


' ---[ Constants/Defines ]------------------------
'
Define LOADER_USED 1      'Only required if bootloader used to
                         'program PIC

' Define ADCIN parameters
Define ADC_BITS      8       ' Set number of bits in result
Define ADC_CLOCK     3       ' Set clock source (3=rc)
Define ADC_SAMPLEUS  50      ' Set sampling time in uS



' ---[ Variables ]------------------------
'
adval  var    byte           ' Create adval to store result

' ---[ Initialization ]------------------------
'
Init:
        PortB = $00          ' all outputs off to start
        TrisB = %00000000    ' All of Port B is outputs


' ---[ Main Code ]------------------------
'

        TRISA = %11111111           ' Set PORTA to all input
        ADCON1 = %00000010    ' Set PORTA analog
```

```
loop:
        ADCIN 2, adval              ' Read channel 0 to adval


'********** Drive LEDs ************************

LEDtst1:
        if adval > 25 then tst2                 'If A/D value is less
                                                'than 25
        portb = %00000001           ' then light LED0 only
        goto cont                   'continue with the program

tst2:
        if adval > 75 then tst3                 'If A/D value is
                                                'between 25 and 75
        portb = %00000011           ' then light LED0 & LED1
        goto cont                   'continue with the program

tst3:
        if adval > 125 then tst4    'If A/D value is between 75
                                    'and 125
        portb = %00000111           ' then light LED0 - LED2
        goto cont                   'continue with the program

tst4:
        if adval > 175 then tst5    'If A/D value is between 125
                                    'and 175
        portb = %00001111           ' then light LED0 - LED3
        goto cont                   'continue with the program

tst5:
        portb = %00011111           'A/D value is greater than 175
                                    'so
                                    ' light all the LEDs 0-4

cont:
        Pause 100                   'wait 1 second
        goto loop


        end
```

Final Thoughts

You can use this concept whenever you want to interface digital circuitry to the real analog world. Reading sensors is probably the most common application but not the only one. The code for this project can easily be turned into a subroutine for more complex programs. You can even modify it to read more than one sensor connected to each of the Port A A/D pins. Don't be discouraged if you have difficulty at first understanding the programs and what they're doing; they'll probably be tough to initially understand because we covered so much in this project.

But now let's change direction again and drive something other than an LED. How about driving a servomotor? The next project does just that.

Project 6—Driving a Servomotor

If you've ever built radio control airplanes or robots, you are probably familiar with *servomotors*. Inside a servomotor is a DC motor with a series of gears attached. The gears drive the output shaft and also control an internal potentiometer. The potentiometer feeds back the output shaft position to the internal control electronics that control the DC motor. The output shaft is limited to 180 degrees of rotation, but some people rework the internals to make the servomotor turn a continuous 360 degrees. (We'll explore applications of such "reworked" servomotors later in this book.) The servomotor is controlled by a pulse-width modulated (PWM) signal. The signal has to be between one and two milliseconds. A 1-millisecond wide pulse moves the shaft all the way to the left, and a 2-millisecond wide pulse moves it all the way to the right. Any pulse width in between moves the shaft between the end points in a linear rotation. A 1.5-millisecond pulse would put the shaft at the halfway point.

This project is quite simple. It first moves the shaft to the center position, and then rotates the shaft back and forth between the end points. It's simple, but quite fun to play with. The finished project is shown in Figure 6-5 and Figure 6-6 shows the schematic diagram.

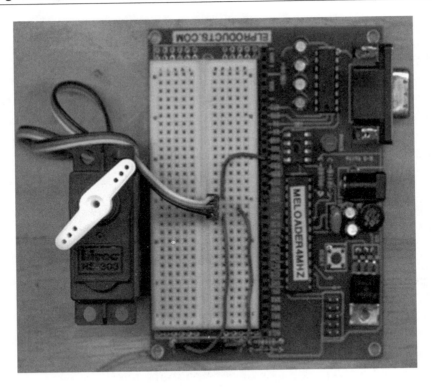

Figure 6-5: Completed servomotor control circuit.

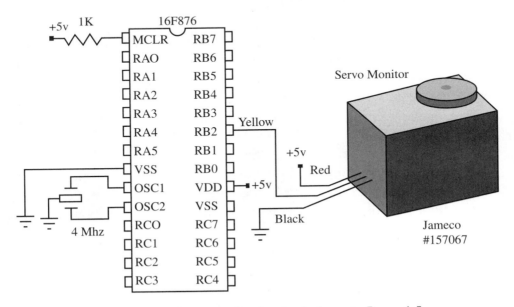

Figure 6-6: Schematic for the circuit shown in Figure 6-5.

The servomotor only requires three wires: 5 volts, ground, and the signal wire that is connected to RB2. Be sure to use a good power supply. The servomotor draws a lot more power than any of the previous projects in this book. If you are using a regulator to produce the 5 volts, I suggest you use at least a TO-220 package with proper heat sinking. Many people use separate power sources for the servomotor and the PIC; just make sure your grounds are all connected if you use that method. The rest of the circuit is the standard PIC connections that were used in the previous projects.

PBC Code

There is really not a lot to explain about the PBC code for this project. It's quite simple. The key command in controlling a servo is the PULSOUT command. (Why they didn't just spell the full word "pulse" in that command is beyond me! I get more syntax errors from spelling it PULSEOUT than any other stupid mistake.) The PULSOUT command requires the PIC pin used to output the signal and the *period* constant to tell how long to send the pulse. That is the format of the command covered earlier in the book.

The PBC code does not require any variables to be established since we are using the pre-defined B0 and B2 byte variables. There is nothing to initialize either. Therefore we jump right into the main code loop. We begin at the Center: label. This block of code centers the shaft of the servomotor. I have the PIC send the PULSOUT signal 100 times by using a FOR-NEXT loop using variable B2. I also have a 20-millisecond pause to allow the servomotor to react. Servomotors require a constant signal to maintain proper position. They have tremendous resistance to movement if you have any loads on the shaft, but they will not hold forever and therefore you should send the position signal often. The 20-millisecond delay is the recommended delay between commands for many servomotors.

The PBC PULSOUT command has a 10-microsecond resolution. The period constant that is used in the center block of code needs to result in a 1.5-millisecond pulse. Therefore the actual value used in the command is 150 (150 * 10 microsecond = 1500 microsecond, or 1.5 milliseconds).

I deliberately send this command 100 times because I found that gave me enough time to pull the linkage arm off the servo and position it back on the motor shaft at center while the motor was being driven to center.

Once the `Center:` loop is complete, the program moves into the `servo:` label block. Here is where the PIC 16F876 is set to drive the servomotor back and forth between the full counter-clockwise range of positions and the full clockwise range of positions. We do this with two FOR-NEXT loops. The first FOR-NEXT loop increments variable B0 by 1, starting with 100 and ending with 200. These are the end points for the servomotor. When the `servo:` label is approached in the code, you will know it because the servo will drive immediately from the center position to most counter-clockwise position. Then the servomotor will slowly step the servo shaft to the full clockwise position. Then the next FOR-NEXT loop increments the servomotor in the opposite direction by changing the B0 variable from 200 to 100 in -1 steps. Notice how the FOR-NEXT command has the `Step -1` statement following it. This is required to make the FOR-NEXT loop count down instead of incrementing B0.

After that second FOR-NEXT is complete, we use a GOTO statement to jump back up to the `servo:` label and do it all again, thus creating a back and forth movement of the servomotor shaft.

The servomotor linkage could be tied to anything you could think of. Ever see one of those Christmas displays where the elf is pounding a hammer? A servomotor could be controlling that with code similar to this program.

```
' —-[ Title ]————————————————--
'
' File...... proj6PBC.bas
' Purpose... PIC 16F876 -> Servo
' Author.... Chuck Hellebuyck
' Started... 15 January 2000
' Updated...


' —-[ Program Description ]————————--
'
'This is a simple program to drive a servo. It initially sets the
'servo to the halfway point of its movement for a short period of
'drives the servo to its full counterclockwise position and then
'to its full
```

```
'clockwise position. The back and forth movement is
'repeated continously in a loop.
'
'Connections:
'PIC-PIN            Servo
'RB2                Control wire (yellow)
'Vdd - 5V           Power Wire (red)
'Vss - Ground       Ground wire (black)
'MCLR - 5V (thru 1k)


' ---[ Constants/Defines ]----------------------------


' ---[ Variables ]----------------------------------


' ---[ Initialization ]---------------------------


' ---[ Main Loop ]-------------------------------


Center:
        For b2 = 1 to 100      'Send center signal 50 times
        pulsout 2, 150         '150 * 10usec = 1.5 msec
        pause 20               'wait 20 msec
        next                   'if 50 are complete move on

servo:
'-----Clockwise Direction----------
        for b0 = 100 to 200    'Move from left to right
        pulsout    2,b0         'send position signal
        pause 20               'wait 20 msec
        next                   'if all positions complete move on
'-----Counter Clockwise Direction-----
        for b0 = 200 to 100 step -1  'move right to left
        pulsout    2,b0               'send position signal
        pause 20                     'wait 20 msec
        next                         'if all positions complete
move on

        goto servo             'loop to servo label and do it again
```

PBPro Code

For this project, the PBC code and PBPro code are almost identical. In fact, the PBPro code is a few lines longer. (I'll repeat much of what I said in the PBC section above in case you're only reading the PBPro sections.)

This code has to first establish the variables. B0 and B2 are not pre-defined. I could have put one of the DEFINE options at the top called:

```
Include "bs1defs.bas"
```

This would establish all the B and W variables the Stamp and PBC use but I don't recommend it. You should always define your variables with the VAR directive. This is a good habit and allows you to name the variables anything you want. That's what I do in the variables section of the code listing. B0 and B2 are established as byte variables. I also have the DEFINE statement for the bootloader added which PBPro requires (PBC never requires this).

After that the code is identical to the PBC version. We use the same PULSOUT command, which requires the PIC pin used to output the signal and the *period* constant to tell how long to send the pulse.

After establishing the variables we jump right into the main code loop. We begin at the Center: label. This block of code centers the shaft of the servomotor. I have the PIC send the PULSOUT signal 100 times by using a FOR-NEXT loop using variable B2. I also have a 20-millisecond pause to allow the servomotor to react. Servomotors require a constant signal to maintain proper position. They have tremendous resistance to movement if you have any loads on the shaft, but they will not hold forever. Therefore you should send the position signal often. The 20-millisecond delay is the recommended delay between commands for many servomotors.

The PBPro PULSOUT command has a 10-microsecond resolution with a 4-MHz crystal/resonator running the PIC. The period constant, which is part of the PULSOUT command, used in the center block of code needs to result in a 1.5-millisecond pulse. Therefore, the actual value used in the command is 150 (150 * 10 microsecond = 1500 microsecond, or 1.5 milliseconds). I deliberately send this command 100 times because I found that gave me enough time to pull the linkage arm off the servomotor and position it back on the motor shaft at center while the servomotor was being driven to center.

Once the `Center:` loop is complete, the program moves into the `servo:` label block. Here is where the PIC 16F876 is set to drive the servomotor back and forth between the full counter-clockwise range of position and the full clockwise range of position. We do this with two FOR-NEXT loops.

The first FOR-NEXT loop increments variable B0 by 1 starting with 100 and ending with 200. These are the end points for the servomotor. When the `servo:` label is approached in the code, you will know it because the servomotor will drive immediately from the center position to most counter-clockwise position. Then the servomotor will slowly step the servomotor shaft to the full clockwise position. Then the next FOR-NEXT loop increments the servomotor the opposite way by changing the B0 variable from 200 to 100 in -1 steps. Notice how the FOR-NEXT command has the `Step -1` statement following it. This is required to make the FOR-NEXT loop count down instead of incrementing B0.

After that second FOR-NEXT is complete we use a GOTO statement to jump back up to the `servo:` label and do it all again thus creating a back and forth movement of the servomotor shaft.

```
' —-[ Title ]——————————————————————  —-
'
' File...... proj6PRO.bas
' Purpose... PIC 16F876 -> Servo
' Author.... Chuck Hellebuyck
' Started... 15 January 2000
' Updated...


' —-[ Program Description ]——————————————  —-
'
'This is a simple program to drive a servo. It initially sets the
'servo to the halfway point of its movement for a short period of
'time and then drives the servo to its full counterclockwise
'position and then to its full clockwise position. The back and
'forth movement is repeated continously in a loop.
'
'Connections:
'PIC-PIN                 Servo
'RB2                     Control wire (yellow)
'Vdd - 5V                Power Wire (red)
'Vss - Ground            Ground wire (black)
'MCLR - 5V (thru 1k)
```

```
' —-[ Constants/Defines ]————————————————--
'
Define LOADER_USED 1        'Only required if bootloader used to
                            'program PIC


' —-[ Variables ]————————————————-
'
B2    var    byte              ' Generic Byte
B0  var byte           ' Generic Byte to store Servo position

' —-[ Initialization ]————————————————
'

' —-[ Code ]————————————--


Center:
        For b2 = 1 to 100     'Send center signal 50 times
        pulsout 2, 150        '150 * 10usec = 1.5 msec
        pause 20              'wait 20 msec
        next                 'if 50 are complete move on

servo:
'————-Clockwise Direction————————
        for b0 = 100 to 200   'Move from left to right
        pulsout 2,b0          'send position signal
        pause 20              'wait 20 msec
        next                 'if all positions complete move on
'————-Counter Clockwise Direction————
        for b0 = 200 to 100 step -1  'move right to left
        pulsout 2,b0                 'send position signal
        pause 20                     'wait 20 msec
        next                        'if all positions complete
move on

        goto servo           'loop to servo label and do it again
```

Final Thoughts

You can easily modify the loops to position the servomotor based on a switch input or even use an A/D input to control the servo. The potentiometer circuit in Project #5 could be combined with this code to make a servomotor that follows the movement of the potentiometer. You could use this to control something from a distance with just a few wires connected between the POT and PIC16F876 circuit and the servomotor somewhere else. You could even have the servomotor controlling something inside a climate-controlled chamber while you control the servomotor from outside the chamber.

CHAPTER 7

Communication

In this chapter, we'll use the power of PicBasic to communicate with a PC serial port using RS232 format. We'll also use PicBasic to drive a parallel LCD module (commonly used to display information). The third project of this chapter will combine the first two projects into one by creating a serial LCD module.

Any of these are good learning projects because many of your PicBasic projects will require some way to communicate information—data, time, etc.—to other humans while the PIC is running. It can even be used to display variable data so you can monitor if your program is running correctly.

There's a lot of material in this chapter, so let's get moving!

Project # 7—Driving an LCD Module

One of the first projects I attempted when I started with PicBasic was to drive a LCD module. LCD modules come in various configurations, but 99% of them use the same interface chip, the Hitachi 44870 LCD character driver. This project is really quite simple but forms the basis for all LCD projects you may build in the future. In this project, we will drive a 2 x16 LCD module and display the phrase output by so many simple computer programs: "Hello World". This project will show how the PBC and PBPro compilers differ while performing the same task and show the basic structure for controlling a LCD.

The schematic for this project, shown in Figure 7-1, is simple. The completed project is shown in Figure 7-2. Note the standard MCLR pull-up resistor and 4-MHz resonator connected to OSC1 and OSC2. The easiest I/O to control with PBC is Port B, so we use that here to control the LCD. The LCD can be controlled in an 8-bit mode or a 4-bit mode, which requires eight I/O or four I/O, respectively. Most people want to save I/O so they use the 4-bit mode. We do the same thing here. Port B is connected to the DB4-DB7 of the LCD module. Through these connections all control characters are sent.

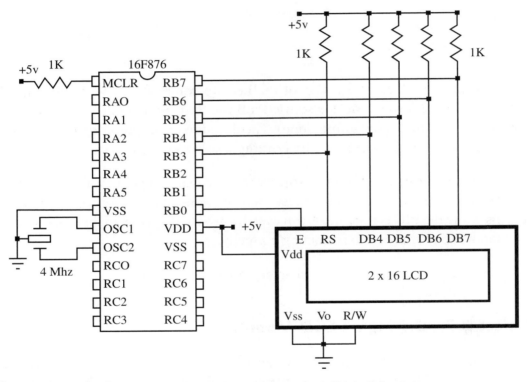

Figure 7-1: Schematic diagram for the LCD module project.

Figure 7-2: View of the complete LCD module circuit board.

The I/O ports have external pull-up resistors to guarantee a logic high level. The PIC Port B has internal pull-up resistors that you can set, but that will cause confusion for the beginning programmer. Therefore, I went with the external resistors.

The LCD has the R/W pin grounded, which limits it to write-only mode, which is all we plan to do anyway. The Vo pin controls the contrast level of the LCD. We ground it for simplicity, and that sets the LCD to maximum contrast. The RS pin is connected to Port B pin 3. It's used to tell the LCD if a character or LCD command is coming from the PIC. The software section will explain this better. Finally +5 volts and ground are connected to the LCD. The LCD pins are not numbered since LCDs come in different pin layouts. Check the data sheet for your LCD to verify the proper pin numbers.

PBC Code

The PBC code has several important steps. First, several variables and constants for the LCD setup are established. Some of them don't get used in this program, but I set them up so you can expand the program without having to add a bunch of new ones. Next, the LCD is initialized through a whole list of commands. The LCD data sheet will explain several steps to set up the LCD. Those steps are spelled out below.

First, we follow the LCD setup process by sending the proper command three times and then all the commands to establish the LCD setup that matches your LCD. The PULSOUT command controls the E, or enable, line of the LCD. The LCD automatically initializes in 8-bit mode. Sending the same command three times and then pulsing the E line for the LCD to recognize it converts the LCD to 4-bit mode.

After that, the LCD is set up with several steps. Notice how almost every other command is a GOSUB to the LCDCMD subroutine. This subroutine controls the RS line of the LCD. That makes the LCD electronics read the information as a command to control the LCD rather than a character to be displayed on the LCD. It does that by setting the RS pin low prior to jumping to the WRLCD subroutine. When the WRLCD subroutine is done, it returns to the LCDCMD subroutine that sets the RS bit back to high and returns to the section of main code that jumped to the LCDCMD subroutine. All these set-up commands are LCD commands that establish the number of rows on the LCD, whether the cursor is on or off, and other minor features.

After the LCD set-up section comes the `main` loop of the code. It starts off by sending commands to clear the LCD. Then it begins to send each character of the phrase "Hello World". Each letter has to be sent separately to the LCD with the same WRLCD subroutine mentioned in the LCD setup section above. Because we are writing characters and not commands, we skip the LCDCMD subroutine and go right to the WRLCD subroutine. The first step is to store the character in the variable "char" and then jump to the WRLCD subroutine.

The WRLCD subroutine is really the heart of the program. Let's look at it in some detail.

```
pins = pins & %00001000 ' output high nibble
```

This line takes all the PORTB pins and does a logical "and" with the binary value %00001000. What that does is reset every bit to 0 except the fourth bit, which is the RS bit. If that bit is a 1 it stays a 1; if it's a 0 it stays a 0. Therefore, it's left untouched because it can sometimes be set by the LCDCMD subroutine that indicates WRLCD is sending a command rather than a character to display.

```
b3 = char & %11110000          'store high nibble of char in B3
```

This next line takes the "char" variable and logically "ANDs" it with binary %11110000. Once again we are clearing the lower four bits but leaving the upper four alone. We do this because we need to break the 8-bit "char" byte into two nibbles so we can send it to the LCD. (Remember we are communicating to the LCD using the 4-bit mode to save I/O.) Notice how we operate on "char" but store the result in b3. This leaves "char" unchanged.

```
pins = pins|b3                 'combine RS signal with char
```

This line combines the PORTB pins with the four unchanged "char" bits to set PORTB with the proper bit states. We do this with the logical "OR" directive. Because of the way we preset the pins and b3 variable, this command just joins the lower four bits of "pins" with the upper four bits of b3 to produce the desired PORTB state.

```
pause 1                        'wait for data setup
```

We pause briefly to let the data stabilize.

```
PULSOUT E, 100            ' strobe the enable line
```

Now we pulse the E line of the LCD so the LCD module knows to accept the data at its data lines. The pulse is a 1-millisecond wide positive pulse.

```
b3 = char * 16                 'Shift low nibble to high nibble
```

This next line is tricky. Since we just sent the upper four bits of the "char" variable, we now have to send the lower four bits. In order to do that, we have to shift the lower four bits to the upper four bits' position. Multiplying any byte by 16 shifts the bits over four places. If we wanted to go the opposite way, we would have divided by 16. Again notice we leave "char" unchanged and store the result in b3.

```
pins = pins & %00001000    'combine RS signal with char
```

Now we reset PORTB I/O back to zeros, except for the RS bit, just like we did at the beginning of this subroutine.

```
pins = pins|b3             ' output low nibble
```

Once again we combine the shifted lower four bits, now in b3, with the lower four bits of PORTB pins data register using the logical "OR" directive.

```
pause 1                    'wait for data setup
```

We pause briefly again to let the data setup.

```
PULSOUT E, 100             'strobe the enable line
```

We pulse the E line so the LCD reads the second set of four bits. The LCD now has the full "char" byte and displays the character on the LCD.

```
RETURN
```

As the final subroutine step, we return back to the area of the program that called the WRLCD subroutine. Actually we return to the command just after the GOSUB WRLCD command line that sent us here.

That's really all there is to this program. After each character of "Hello World" is sent to the LCD, we just pause for a second and then loop back to do it again.

```
' ---[ Title ]--------------------------------------   --
'
' File...... Proj7PBC.BAS
' Purpose... PIC -> LCD (4-bit interface) using 16F876 and 2x16
' LCD
' Author.... Chuck Hellebuyck
' Started... January 20, 2002
' Updated...


' ---[ Program Description ]-----------------------   --
'
'
' PIC16F876 to LCD Port predefined connections:
'
```

```
' PIC              LCD                 Other Connections
' ___              ___-                _____-
' B4                      LCD.11
' B5               LCD.12
' B6               LCD.13
' B7               LCD.14
' B3               LCD.4
' B0               LCD.6
' OSC1                             Resonator - 4 mhz
' OSC2                             Resonator - 4 Mhz
' MCLR                             Vdd via 1k resistor
' Vdd                              5v
' Vss                              Gnd


' ---[ Revision History ]_____
'
'


' ---[ Constants ]_____--
'
' LCD control pins
'
symbol E  =  0      ' LCD enable pin (1 = enabled)
symbol RS = 3       ' Register Select (1 = char)

' LCD control characters
'
symbol ClrLCD  =  $01        ' clear the LCD
symbol CrsrHm  =  $02       ' move cursor to home position
symbol Row2    =  $C0        ' 2nd row position of LCD
symbol CrsrLf  =  $10       ' move cursor left
symbol CrsrRt  =  $14       ' move cursor right
symbol DispLf  =  $18       ' shift displayed chars left
symbol DispRt  =  $1C       ' shift displayed chars right
symbol Digit   =  $30       ' Character column code for LCD
' ---[ Variables ]_____--
'
symbol x = B0             ' General purpose variable
symbol char = B1          ' char sent to LCD
symbol loop1= B2          ' loop counter


' ---[ Initialization ]_____-
'
Init:  pins = $0000     ' all outputs off to start
```

```
        Dirs = %11111111      ' LCD pins
        PAUSE 215             ' pause for LCD setup

' Initialize the LCD (Hitatchi HD44780 controller)
'
I_LCD:
        pins = %00110000     'set to 8 bit operation
        PULSOUT E,100              'SEND DATA 3 TIMES
        PAUSE 10
        PULSOUT E,100             'SEND DATA 3 TIMES
        PAUSE 10
        PULSOUT E,100             'SEND DATA 3 TIMES
        PAUSE 10
        PINS = %00100000     'SET TO 4 BIT OPERATION
        pause 1
        PULSOUT E,100             'SEND DATA 3 TIMES
        HIGH RS
        CHAR = %00101000     '4 BIT, 2 LINES, 5X7 CHARACTER
        GOSUB LCDCMD
        CHAR = 8             'Display, cursor and blink off
        GOSUB LCDCMD
        CHAR = 6             'Shift display right
        GOSUB LCDCMD
        CHAR = 1             'clear display and return home
        GOSUB LCDCMD
        CHAR = 15            'display, cursor and blink on
        GOSUB LCDCMD


' —-[ Main Loop ]——————————————————--
'

Start:char = clrlcd       ' Clear LCD
        gosub lcdcmd                ' and position cursor at home
        char = CRSRHM                ' Issue cursor home command
        gosub lcdcmd                ' send cursor home on LCD

'***** Send "Hello World" to first line of LCD ******
        char = "H"                      ' Send "Hello World" one
                                        ' letter
        gosub wrlcd                     ' at a time to the LCD
        char = "e"
        gosub wrlcd
        char = "l"
        gosub wrlcd
        char = "l"
```

```
          gosub wrlcd
          char = "o"
          gosub wrlcd
          char = " "
          gosub wrlcd
          char = "W"
          gosub wrlcd
          char = "o"
          gosub wrlcd
          char = "r"
          gosub wrlcd
          char = "l"
          gosub wrlcd
          char = "d"
          gosub wrlcd

          Pause 1000            'Pause long enough to see it

          goto start

' Send command byte to LCD Subroutine
'
LCDcmd:
          LOW RS                ' RS low = command
          GOSUB WrLCD           ' send the byte
          HIGH RS               ' return to character mode
          RETURN

' Write ASCII char to LCD Subroutine
'
WrLCD:
      pins = pins & %00001000   ' output high nibble
          b3 = char & %11110000 'store high nibble of char in B3
          pins = pins|b3        'combine RS signal with char
          pause 1               'wait for data setup
          PULSOUT E, 100        ' strobe the enable line
          b3 = char * 16        'Shift low nibble to high nibble
          pins = pins & %00001000  'combine RS signal with char
          pins = pins|b3        ' output low nibble
          pause 1               'wait for data setup
          PULSOUT E, 100        'strobe the enable line
          RETURN
```

PBPro Code

The PBPro version of the code demonstrates one of PBPro's major advantages over PBC. All of that set-up routine we did in the PBC example, and all of the LCDCMD and WRLCD subroutine stuff, is done within the PBPro command LCDOUT. In fact, we still have control over the LCD set-up, but we do it with DEFINE statements rather than a series of GOSUB commands.

The first part of the program establishes all the DEFINE statements to tell PBPro what port to use for the data port, RS line, and E line. Each define indicates which pin(s) of the port are for communication. We also use DEFINE to communicate the 4-bit mode setup and the number of LCD lines. Finally we even have a DEFINE to control the time between commands being sent and time delay for data set-up. Some LCDs are picky, so PBPro allows you to adjust the timing of its LCDOUT command to work with various LCDs. The DEFINE statements I used here should work with most LCDs since I really slowed things down.

```
lcdout $fe, 1                          ' Clear LCD
```

The main loop is very small but does the same thing as the longer PBC program we previously examined. The LCDOUT command line above is sending a command to the LCD to clear the screen. The LCDOUT command is sending it as a command signal because of the "$fe" in front of the code "1" for clearing the LCD. All the toggling of the RS line is done by the LCDOUT command. All you have to remember is to put the "$fe" in front of the code so PBPro knows you meant to send a command.

```
LCDOUT   "Hello World"        ' Send "Hello World"  to the LCD
```

The next line sends the characters within the quote marks, "Hello World". The LCDOUT command allows you to put the whole phrase between quotes and then it sends it, character by character, to the LCD. The "$fe" is left off because these are characters to display and not command codes. After this line, we pause for 1 second and loop around to do it again.

See how much easier PBPro is to use? The LCDOUT command is priceless in my opinion and is one of the reasons PBPro is worth the extra money it costs. Even though the actual compiled code is not a lot smaller than the PBC code, the listing is much smaller and easier to read.

```
' —-[ Title ]——————————————————--
'
' File...... proj7pro.BAS
' Purpose... PIC -> LCD (4-bit interface) using 16F876 and 2x16
' LCD
' Author.... Chuck Hellebuyck
' Started... November 19, 1999
' Updated...


' —-[ Program Description ]————————————--
'
'
' PIC16F876 to LCD Port predefined connections:
'
' PIC            LCD           Other Connections
' ___            ___-          _____-
' B4          LCD.11
' B5          LCD.12
' B6          LCD.13
' B7          LCD.14
' B3          LCD.4
' B0          LCD.6
' OSC1                      Resonator - 4 mhz
' OSC2                      Resonator - 4 Mhz
' MCLR              Vdd via 1k resistor
' Vdd                       5v
' Vss                       Gnd


' —-[ Revision History ]————————————
'
'
' —[ Includes / Defines ]————————————--
'
Define LOADER_USED 1      'Only required if bootloader used to
                         'program PIC

DEFINE LCD_DREG    PORTB  'Define PIC port used for LCD Data
                         'lines
DEFINE LCD_DBIT    4         'Define first pin of portb
                              'connected to LCD DB4
DEFINE LCD_RSREG PORTB    'Define PIC port used for RS line of
                         'LCD
DEFINE LCD_RSBIT 3        'Define Portb pin used for RS
                         'connection
```

```
DEFINE LCD_EREG      PORTB   'Define PIC prot used for E line of LCD
DEFINE LCD_EBIT      0               'Define PortB pin used for E
                                     'connection
DEFINE LCD_BITS      4               'Define the 4 bit communication
                                     'mode to LCD
DEFINE LCD_LINES 2           'Define using a 2 line LCD
DEFINE LCD_COMMANDUS 2000 'Define delay between sending LCD
                             ' commands
DEFINE LCD_DATAUS 50         'Define delay time between data sent.

' —-[ Constants ]————————————————-
'
' —-[ Variables ]————————————————-
'
' —-[ Initialization ]——————————————
'
' —-[ Main Code ]————————————————-
'

Start:
       lcdout $fe, 1                 ' Clear LCD
       lcdout $fe, 2                 ' Position cursor at home


'***** Send "Hello World" to first line of LCD ******

    LCDOUT  "Hello World"  ' Send "Hello World"  to the LCD
    Pause   1000                     ' Pause for 1 second to see it

    Goto Start                       ' Loop back and do it all
                                     ' again
```

Final Thoughts

As this project illustrates, PBC can do a lot, but PBPro can do more. Imagine if you had several messages you wanted to display on the LCD. With PBC, you would have to spell out each character with a separate command. You could set up a lookup table and send it from a loop. That would save code space, but PBPro would just require a single LCDOUT command for each displayed message.

These two programs can be easily modified to fit into any PicBasic program you write to have an LCD display as part of the project.

Project #8—Serial Communication

This project requires a terminal program running on your PC. The editor software I'm using is called "Codestudio" and has a built-in terminal window. Many of the windows interface software programs available as shareware on the web for using PBC and PBPro have a terminal window. In this project, we will communicate with the PIC16F876 using the serial port of the PC and display info in that terminal window. This whole project is built around the SERIN and SEROUT commands. Those commands can be used on any PIC pin.

The 16F876 has a dedicated serial buffer built in on pins C6 and C7. We won't use those here since I wanted to demonstrate the versatility of the SERIN and SEROUT commands. We will use a serial buffer chip that shifts the 0–5 volts PIC signal to the −12 v to +12 v signal the PC serial port likes to see. Some PCs will read the 0–5 volt signal, but it's safer to buffer your circuit from the PC serial port.

This project will send a menu of commands to the PC to be displayed in the terminal window. You will choose from that menu and send back your selection from the terminal window, through the serial port, and back to the PIC circuit. Based on which selection you make at the PC, the PIC will respond with a message or change the state of an LED. From this you will easily see how we could control a PIC-based module directly from a PC serial port.

Figure 7-3 shows the schematic for this project while Figure 7-4 shows the completed circuit. The standard resonator, MCLR, and power/ground connections are present as in previous projects. Added are the connections to the RS232 level shifter chip and DB9 connector that will hook to a PC "straight-thru" serial cable. We will use that standard cable rather than a null-modem cable. The level shifter chip is available from various sources and all have the same pin-out. The circuit also has an LED connected to Port B pin 0. We use this as a visual indicator that is controlled from the PC. By choosing the proper selection from the menu, you can control the state of the LED.

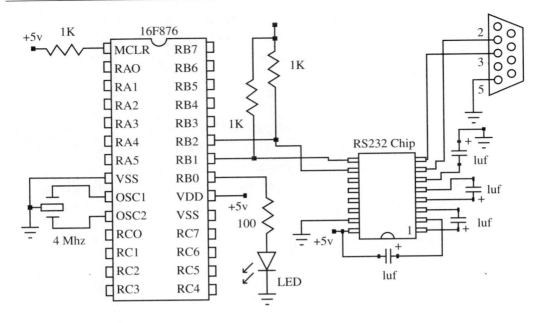

Figure 7-3: Schematic diagram for the serial communications project.

The PBC code starts off initializing the LED to the off state. (This shouldn't take much explanation by now!) The main part of the program at label `Menu` creates the menu you will see on the PC screen terminal window. It does this by sending each line of the menu serially to the terminal program using the `SEROUT` command. The program uses 2400 baud true mode to communicate. The terminal program has to be set at the same baud rate. We have to use the T2400 true mode directive so the RS232 level shifter chip sees the proper signal levels.

The information between quotes in the `SEROUT` command lines are sent as ASCII bytes. The PC should recognize the characters and display the word "Menu" as the first line. For each number displayed as part of the menu, we place a "#" symbol in front of the actual number so the ASCII equivalent is sent instead of just the number. You see, if we just sent the number "1" the ASCII character associated with the value one would be displayed. Instead we want the ASCII character associated with hex $31, which is the ASCII character "1". By putting a "#" in front of the number, PBC will send a $31 instead of a $01.

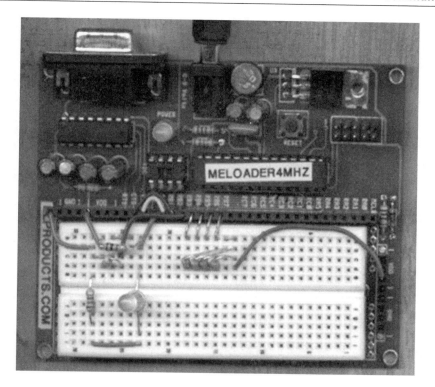

Figure 7-4: Completed circuit for the serial communications project.

Each SEROUT command line ends with sending the 10 and 13 characters. These are the ASCII codes for line feed and carriage return. These make each line of SEROUT sent information display as a separate line on the PC. It might help you to look up the ASCII character set to understand all the characters and their code. You can find that in Appendix B.

The next section at label receive is where the PIC waits for the menu choice to be sent by the PC. The choice is sent as an ASCII value. To convert that back into a decimal number we can use, we subtract $30 (30 hex). All ASCII numbers are offset by $30 (0 = $30, 1 = $31, etc). Once we have the menu selection as a numeric value, we can use that to operate on the user's choice. We use that numeric value to branch to one of four locations. They are labeled Zero, One, Two, and Three. At each label we have a different function; some are simply a single line that sends back a "Hello" or "Goodbye".

At label One and label Two, we simply send back serial "Hello" or "Goodbye" and then return to the menu routine to redisplay the menu choices. At label Three, we have the PIC control a LED. Each time you enter choice three at the PC, the LED connected to the PIC Port B pin 0 reverses its state from on to off or off to on. We can tell what state it was in previously by the bit flag "LED." If it's a 0, then we know the LED is off, so go to the routine that turns the LED on. If it's a 1, then we go to the routine that turns the LED off. We also send a message with the state of the LED using the SEROUT command. After that we return to the menu routine to display the menu choices again. That's really all there is to it.

```
' ---[ Title ]------------------------------ --
'
' File...... proj8pbc.BAS
' Purpose... PIC -> PC serial port using 16F876
' Author.... Chuck Hellebuyck
' Started... November 9, 2001
' Updated...


' ---[ Program Description ]----------------- --
'
'
' PIC16F876 hardware connections
' PIC           External
' ---          -------
' RB1           Max232(RX)
' RB2           Max232(TX)
' RB0           LED
' MCLR          5v
' Vdd           5v
' Vss           gnd
' Gnd           DB9-pin 5 (gnd)


' ---[ Revision History ]-------------------
'
'

' ---[ Constants ]-------------------------- --
'

' ---[ Variables ]-------------------------- --
'
symbol RX = B2              ' Receive byte
```

```
symbol LED = bit0    ' LED status bit


' ——-[ Initialization ]————————————————————
'
Init:
LED = 0                      'Initialize LED flag to zero
low 0                    'Initialize LED to off


' ——-[ Main Code ]———————————————————————-
'
Menu:
' ****** [Menu setup on PC screen] ************************
     serout 2, T2400, ("menu", 10, 13)              'Display menu
                                                    'on PC screen
     serout 2, T2400, (#1, ") ", "send hello", 10, 13)
     serout 2, T2400, (#2, ") ", "send goodbye", 10, 13)
     serout 2, T2400, (#3, ") ", "toggle LED", 10, 13)

Receive:
' ***** [Receive the menu selection from PC] **************
     serin 1, T2400, RX          'Receive menu number
     RX = RX - $30                       'Convert ASCII number to
                                         ' decimal
     If RX > 3 then error                'Test for good value
     Branch RX, (zero, one, two, three)      'redirect to menu
                                             'selection code
Error:
     serout 2, T2400, ("error", 10, 13, "Try again", 10, 13)
     goto menu

Zero:
'***** [ Code for zero value ] ****************************
     goto menu                   'Return to menu, zero is not a
                                 'valid selection

One:
'***** [Code for selection 1] ************************
     serout 2, T2400, ("Hello",13,10,13)    'Send "Hello" back
                                            'to PC
     goto menu                              'Return to main menu
                                            'routine

Two:
'***** [Code for selection 2] ***********************
```

```
        serout 2, T2400, ("Goodbye",13,10,13)     'Send "Goodbye" to
                                                   'PC
        goto menu                          'Return to Menu routine

Three:
'*****  [Code for selection 3]  ********************
        if LED = 1 then off                    'If LED bit =1 then goto
                                               'off
        high 0                             'Turn LED on
        led = 1                                    'Set LED bit to 1
        serout 2, T2400, ("LED ON",13,10,13)   'Send LED status to
                                               'PC
        goto menu                          'Return to main menu

Off:
        low 0                              'Turn LED off
        led = 0                                'Clear LED bit to 0
        serout 2, T2400, ("LED OFF",13,10,13)  'Send LED status to
                                               'PC
        goto menu                          'Return to main menu

Goto menu
```

PBPro Code

The PBPro code starts off initializing the LED to the off state. PBPro also requires a DEFINE to establish the SEROUT mode definitions. PBPro doesn't automatically recognize the T2400 mode in the SEROUT command without the MODEDEFS.BAS INCLUDE. PBPro also has to set up the TRISB and PORTB registers so the proper state of the pins is established.

The main part of the program at label Menu sets up the menu you will see on the PC screen terminal window. It does this by sending each line of the menu serially to the terminal program using the SEROUT command. The program uses 2400-baud true mode to communicate. The terminal program has to be set up at the same baud rate. We have to use the T2400 true mode directive so the RS232 level shifter chip sees the proper signal levels.

The information between quotes in the SEROUT command lines is sent as ASCII bytes. The PC should recognize the characters and display the word "Menu" as the first line. For each number displayed as part of the menu, we place a "#" symbol in front of the actual number so the ASCII equivalent is sent instead of just the number. If we just sent the number "1" the ASCII character associated with the value one would be displayed. Instead we want the ASCII character associated with hex $31, which is the ASCII character "1". By putting a "#" in front of the number, PBPro will send a $31 instead of a $01.

Each SEROUT command line ends with sending the 10 and 13 characters. These are the ASCII codes for line feed and carriage return. These make each line of SEROUT send information and display as a separate line on the PC. Again, it might help you to look up the ASCII character set to understand all the characters and their code. You can find it in Appendix B.

The next section label receive is where the PIC waits for the menu choice to be sent by the PC. The choice is sent as an ASCII value. To convert that back into a decimal number we can use, we subtract $30 (30 hex). All numbers are offset by $30 (0 = $30, 1 = $31, etc). Once we have the menu selection as a numeric value, we can use that to operate on the user's choice. We use that numeric value to branch to one of four locations. They are labeled Zero, One, Two, and Three. At each label, we have a different function. Some are simply a single line that sends back a "Hello" or "Goodbye". ·

At label One and label Two we simply send back serial "Hello" or "Goodbye" and then return to the menu routine to redisplay the menu choices. At label Three, we have the PIC control an LED. Each time you enter choice three at the PC, the LED connected to the PIC Port B pin 0 reverses its state from on to off, or off to on. We can tell what state it was in previously by the bit flag "LED." If it's a 0, then we know the LED is off so go to the routine that turns the LED on. If it's a 1, then we go to the routine that turns the LED off. We also send a message with the state of the LED using the SEROUT command. After that we return to the menu routine to display the menu choices again.

```
' —-[ Title ]——————————————————————————--
'
' File...... proj8pro.BAS
' Purpose... PIC -> PC serial port using 16F876
' Author.... Chuck Hellebuyck
' Started... November 9, 2001
' Updated...
```

```
' —-[ Program Description ]——————————————--
'
'
' PIC16F876 hardware connections
' PIC         External
' —           ——-
' RB1         Max232(RX)
' RB2         Max232(TX)
' RB0         LED
' MCLR        5v
' Vdd         5v
' Vss         gnd
' Gnd         DB9-pin 5 (gnd)


' —-[ Revision History ]————————————————
'
'
' —-[ Includes/Defines ]————————————————--
'
include "modedefs.bas"      'include serout defines
define loader_used 1              'Used for bootloader only

' —-[ Constants ]————————————————————--
'

' —-[ Variables ]————————————————————--
'
RX var byte          ' Receive byte
LED var bit          ' LED status bit


' —-[ Initialization ]————————————————
'
Init:
TRISB = %00000010           'All port b output except pin 1 (RX) is
                            'input
PORTB = %00000000           'Initialize PortB to all zeros and LED
                            'to off

LED = 0                     'Initialize LED flag to 0


' —-[ Main Code ]————————————————————--
'
Menu:
```

```
' ****** [Menu setup on PC screen] *************************
      serout 2, T2400, ["menu", 10, 13]              'Display menu
                                                      'on PC screen
      serout 2, T2400, [#1, ") ", "send hello", 10, 13]
      serout 2, T2400, [#2, ") ", "send goodbye", 10, 13]
      serout 2, T2400, [#3, ") ", "toggle LED", 10, 13]

Receive:
' ***** [Receive the menu selection from PC] **************
            serin 1, T2400, RX                  'Receive menu
                                                 'number
      RX = RX - $30                              'Convert ASCII
                                                 'number to decimal
            If RX > 3 then Error                  ' Test for
                                                 'good value
      Branch RX, [zero, one, two, three]         'redirect to menu
                                                 'selection code
Error:
      serout 2, T2400, ["error", 10, 13, "Try again", 10, 13]
      goto menu

Zero:
'***** [ Code for zero value ] ***************************
      goto menu                   'Return to menu, zero is not a
                                  'valid selection

One:
'***** [Code for selection 1] ************************
      serout 2, T2400, ["Hello", 10, 13]      'Send "Hello" back
                                              'to PC
      goto menu                               'Return to main menu
                                              'routine

Two:
'***** [Code for selection 2] ************************
      serout 2, T2400, ["Goodbye", 10, 13]    'Send "Goodbye" to
                                              'PC
      goto menu                               'Return to Menu routine

Three:
'***** [Code for selection 3] ********************
      if LED = 1 then LEDoff                  'If LED bit =1 then
                                              'goto off
      portb.0 = 1                             'Turn LED on
      led = 1                                 'Set LED bit to 1
```

```
        serout 2, T2400, ["LED ON", 10, 13]      'Send LED status to
                                                  'PC
        goto menu                                 'Return to main menu

LEDOff:
        portb.0 = 0                               'Turn LED off
        led = 0                                   'Clear LED bit to 0
        serout 2, T2400, ["LED OFF", 10, 13]      'Send LED status to
                                                  'PC
        goto menu                                 'Return to main menu

Goto menu
```

Final Thoughts

As you can see, the PBC and PBPro programs are very similar for this project. You can easily expand the menu to include more choices. You can also have the menu choices do a lot more than light a LED or send back a message; for example, you could have a series of control circuits tied to the PIC pins and control them through this same setup. Or how about a robot arm in a "cold chamber"? You could have that robot arm controlled by a PIC and that PIC controlled through a single serial connection to a PC in a warm lab. Interesting?

Project #9—Driving an LCD with a Single Serial Connection

This project uses the same hardware connections as Projects #7 and #8, but the software is unique. The idea is to receive information through the serial port and then display the information on the LCD. This allows a serial port from any PC or another PIC to drive the LCD with a single serial connection.

The software works on simple principles. The PIC waits for three bytes of data. The first is the "row" byte. It indicates on which row the information should be displayed.

The second byte is the "location" byte. It indicates at which position (column) within the row the information should start. The third and final byte is the character code of the information to be displayed. This would be the code for a letter ("a") or number ("1"). This is defined by the display character generator, but is typically an ASCII value similar to what we did in Project #8 to set up the PC menu.

The third byte has an alter ego, though; it can also be used to control the LCD via custom code commands. To enter this command control mode, we set the "row" byte equal to 0. When the module receives the 0 value for the "row" byte, then the module knows that the character code is a command code and not a character to be displayed. A separate action routine will occur based on that command code. Clearing the whole display would be one such command code.

The schematic for this project is shown in Figure 7-5; it combines the schematics of Projects #7 and #8 into one schematic. It's not too different from those projects, so there is not much to explain here. We use the same resonator and MCLR pull-up resistor as every other project. The serial connection is the same as Project #8 and the LCD connection is the same as Project #7. The completed circuit is shown in Figure 7-6.

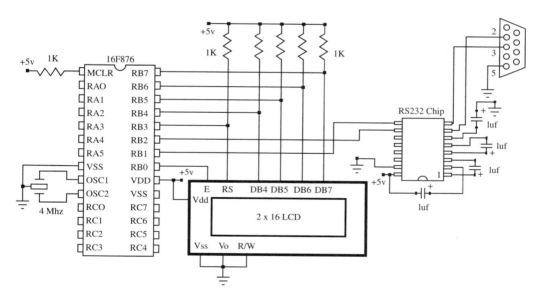

Figure 7-5: Schematic diagram for driving a LCD through a serial connection.

Figure 7-6: Finished circuit board for the circuit in Figure 7-5.

PBC Code

While the circuit for this project may be simple, the PBCcode for it may initially look confusing. Let's break it down.

The labels `Init` and `I_LCD` initialize the LCD the same way Project #7 did. We use the 2x16 module because those are very common LCD modules. You can reconfigure the code to work with any LCD size by first modifying the initialization section.

The next label, `start`, is where this project code really starts. The main line is the `SERIN` line that waits for the row, location, and value bytes. The program will sit here forever if it doesn't receive any information. When information is received, it first tests the row byte and location byte to see if "Row" is not 0 and "Location" is

0, indicating the cursor should not be moved from its existing position and to write the character in the value byte at that location. It does that by jumping to the Display label.

If the location byte is not 0, independent of the value of "Row," then the next command that is run which is a BRANCH instruction. The BRANCH instruction jumps the program to the proper label based on the value of the row byte. If the row byte is 0, then the program jumps to the label Command. The code at the Command label will send a LCD command based on the value byte received (I'll explain this in more detail later).

If "Row" does not equal 0, then the BRANCH command redirects the program to the R1 or R2 labels. Let's try to follow that path. At labels R1 and R2, the program converts the location byte received into the corresponding LCD code to position the LCD cursor at the proper row and column using the LOOKUP command. We have to subtract 1 from the location because the LOOKUP command starts at 0 instead of 1.

After we have the proper position code from the LOOKUP command, the program then jumps to the LCDcmd subroutine to send the special LCD position code to the LCD and move the cursor on the LCD. When the subroutine is done, the program jumps back to the command after the GOSUB LCDcmd line. That command is a jump to the Display label.

At the Display label, the program first stores the serial-received "Value" byte in the "Char" variable. Then the program jumps to the subroutine WrLCD to send that byte to the LCD character generator that actually displays the character on the LCD. The program then returns back to the Display label routine, which then sends the program back to the top to receive a new set of information at the Start label. (See how we first positioned the cursor based on the row and location bytes and then sent the character code to be displayed at that position?)

If the "Row" byte is 0, the BRANCH command under the Start label redirects the program to the Command label. We don't stop at the R1 and R2 labels to get a LCD command code because the "Value" byte received serially should have the LCD command byte in it. All we have to do is convert that value byte to the proper LCD command code based on the table above the Command label. First we take the "Value" byte and convert it into a decimal number by subtracting hex $30. Then we use the LOOKUP command to change the "Value" byte into the proper LCD command code. That command code is stored in the "Char" variable.

The next command line jumps the program to the LCDcmd subroutine where the command code is sent to the LCD. After that is complete we return from the subroutine and then jump back to the Start label to receive more data.

That's really all this program does. It hopefully was clear to you, as this program jumps around a lot.

```
' —-[ Title ]————————————————————--
'
' File...... proj9pbc.BAS
' Purpose... Serial -> PIC16F876 -> LCD (4-bit interface)
' Author.... Chuck Hellebuyck
' Started... January 20,2002
' Updated...


' —-[ Program Description ]————————————--
'
'
' PIC16F876 Port Hardware connections:
'
' PIC              LCD                Other Connections
' ———             ———--               ————————————--
' B4              LCD.11
' B5              LCD.12
' B6              LCD.13
' B7              LCD.14
' B3              LCD.4
' B0              LCD.6
' OSC1                               Resonator - 4 mhz
' OSC2                               Resonator - 4 Mhz
' MCLR                    Vdd via 1k resistor
' Vdd                     5v
' Vss                     Gnd
' B1                                     Max232(RX)
' B2                                     Max232(TX)

' —-[ Revision History ]————————————
'
'
```

```
' —-[ Constants ]————————————————————————-

'
' LCD control pins
'
symbol E  =  0                          ' LCD enable pin (1 = enabled)
symbol RS = 3                           ' Register Select (1 = char,
                                        ' 0 = command)

' LCD control characters
'
symbol ClrLCD   =   $01                 ' clear the LCD
symbol CrsrHm   =   $02                 ' move cursor to home position
symbol Row2     =   $C0                 ' 2nd row position of LCD
symbol Row3 = $94                       ' 3rd row position of LCD
symbol Row4 = $D4                       ' 4th row position of LCD
symbol CrsrLf   =   $10                 ' move cursor left
symbol CrsrRt   =   $14                 ' move cursor right
symbol DispLf   =   $18                 ' shift displayed chars left
symbol DispRt   =   $1C                 ' shift displayed chars right
symbol Digit    =   $30                 ' Character column code for LCD
' —-[ Variables ]————————————————————————--
'
'B3 reserved for drive routine
symbol x = B0                           ' General purpose variable
symbol char = B1                    ' char sent to LCD
symbol loop1= B2                    ' loop counter
symbol ROW = b5                      ' LCD ROW value
symbol LOCATION = b6                ' Column position on the LCD
symbol VALUE = b7                   ' Value is the ASCII Character to
                                    ' display
symbol temp2 = b8                   ' unused
symbol temp3 = b9                   ' unused
symbol temp4 = b10                  ' unused
symbol temp1 = b11                  ' unused


' —-[ Initialization ]—————————————————————
'
Init:
        pins = $0000                     ' all outputs off to start
        Dirs = %11111101             ' LCD pins
        PAUSE 215                                ' pause for LCD
                                             ' setup


' Initialize the LCD (Hitatchi HD44780 controller)
'
I_LCD:
```

```
        pins = %00110000          'set to 8 bit operation
        PULSOUT E,100             'SEND DATA 3 TIMES
        PAUSE 10
        PULSOUT E,100             'SEND DATA 3 TIMES
        PAUSE 10
        PULSOUT E,100             'SEND DATA 3 TIMES
        PAUSE 10
        PINS = %00100000          'SET TO 4 BIT OPERATION
        pause 1
        PULSOUT E,100             'SEND DATA 3 TIMES
        HIGH RS
        CHAR = %00101000          '4 BIT, 2 LINES, 5X7 CHARACTER
        GOSUB LCDCMD
        CHAR = 8                  'Display, cursor and blink off
        GOSUB LCDCMD
        CHAR = 6                  'Shift display right
        GOSUB LCDCMD
        CHAR = 1                  'clear display and return home
        GOSUB LCDCMD
        CHAR = 15                 'display, cursor and blink on
        GOSUB LCDCMD


' —-[ Main Code ]————————————————————————————————--
'
'Display initial screen
'
'******* Main Program ****************************************

Start:
        SERIN 1,T2400,ROW,LOCATION,VALUE 'receive serial data
        Row = Row - $30                  ' Correct row to decimal
                                         ' number
        Location = Location - $30        ' Correct location to
                                         ' decimal number


'******* Decision and Branch Routine ****************************
' if ROW = 0 then its a command so jump to command
' if ROW does not = 0 and LOCATION = 0 then write the value where
' the cursor is. VALUE is the ASCII equivalent of the value sent
' per the LCD ASCII chart

        IF ROW <> 0 AND LOCATION = 0 THEN DISPLAY    'Test for
                                                     'zero value
        BRANCH ROW,(COMMAND,R1,R2)',R3,R4)           'Branch to
                                                     'proper row
```

```
'******* position cursor where new value will be written *********

R1:
        LOCATION = LOCATION - 1      'Correct location byte for zero
                                     'value
' Shrunk to fit in one line, lookup table for cursor position LCD
' code
LOOKUP LOCATION, ($80,$81,$82,$83,$84,$85,$86,$87,$88,$89,$8A,$8B,
    $8C,$8D,$8E,$8F,$90,$91,$92,$93),char
        GOSUB LCDcmd                 'Send position command to LCD
        GOTO DISPLAY                 'Go to character display
                                     'routine

R2:
        LOCATION = LOCATION - 1      'Correct location byte for zero
                                     'value
'Shrunk to fit in one line, lookup table for cursor position LCD
'code
LOOKUP LOCATION, ($C0,$C1,$C2,$C3,$C4,$C5,$C6,$C7,$C8,$C9,$CA,$CB,
    $CC,$CD,$CE,$CF,$D0,$D1,$D2,$D3),char
        GOSUB LCDcmd                 'Send position command to LCD
        GOTO DISPLAY                 ' Go to character display
                                     ' routine


'**** row 3 and row 4 setup for 4x16 LCD. These command commented
'out.
'R3:
'        LOCATION = LOCATION - 1
'LOOKUP LOCATION, ($94,$95,$96,$97,$98,$99,$9A,$9B,$9C,$9D,$9E,$9F,
'$A0,$A1,$A2,$A3,$A4,$A5,$A6,$A7),char
'               GOSUB LCDcmd
'        GOTO DISPLAY
'
'R4:
'        LOCATION = LOCATION - 1
'LOOKUP LOCATION, ($D4,$D5,$D6,$D7,$D8,$D9,$DA,$DB,$DC,$DD,$DE,$DF,
'$E0,$E1,$E2,$E3,$E4,$E5,$E6,$E7),char
'        GOSUB LCDcmd
'        GOTO DISPLAY
```

```
'****** convert value to be displayed for wrlcd routine
****************

DISPLAY:
        char = VALUE            ' Store received Value character in
                               ' char variable
        GOSUB WrLCD            ' Jump to routine that sends char to
                               ' LCD
        GOTO START            ' jump back to beginning of main loop

'***** ROW=0, Therefore run a command per list below
*******************

'0      clear LCD and move to position 1 of row 1
'1      shift cursor left
'2      Display is off, Cursor is off, Cursor Blink is off
'3      Display is on, Cursor is off, Cursor Blink is off
'4      Display is on, Cursor is on, Cursor Blink is off
'5      Display is on, Cursor is off, Cursor Blink is on
'6      Display is on, Cursor is on, Cursor Blink is on
'7      shift display right
'8      shift display left
'9      shift cursor right

COMMAND:
        value = value - $30         'Correct value to decimal for
                                   'command byte only
        '*** Convert byte received to LCD command code
        LOOKUP VALUE,($01,$10,$08,$0C,$0E,$0D,$0F,$1C,$18,$14),char
        GOSUB LCDcmd            'Jump to routine that sends command to
                               'LCD
        GOTO START            'Jump to beginning of main loop

' ****** Send command byte to LCD Routine *************
'LCDcmd:
        LOW RS                          ' RS low = command
        GOSUB WrLCD                  ' send the command byte
        HIGH RS                          ' return to character mode
        RETURN                          ' Return to caller of this
                                       ' subroutine
```

```
'******* Write ASCII char to LCD  **********
WrLCD:
        pins = pins & %00001000   ' Set LCD data lines to zero,
                                  ' leave RS bit alone
        b3 = char & %11110000     'store high nibble of char in B3
        pins = pins|b3                  'Output high nibble and RS
                                        'signal
        pause 1                   'Wait for data setup
        PULSOUT E, 100            'Strobe the enable line
        b3 = char * 16                  'Shift low nibble to high
                                        'nibble
        pins = pins & %00001000   'Set LCD data lines to zero,
                                  'leave RS bit alone
           pins = pins|b3                       'Output low
                                                'nibble and RS
                                                'signal
           pause 1                      'Wait for data setup
        PULSOUT E, 100              'Strobe the enable line
        RETURN               ' Return to where this subroutine was
                             ' called

        END
```

PBPro Code

The PBPro code also may initially look confusing. The first part of the program establishes all the DEFINE statements to tell PBPro which port to use for the data port, RS line, and E line. Each DEFINE then directs which pin(s) of the port are for communication. We even use the DEFINE statements to communicate the 4-bit mode and the number of LCD lines. Finally, we even have a DEFINE to control the time between commands being sent and time delay for data set-up. Some LCDs are picky so PBPro allows you to adjust the timing of its LCDOUT command to work with various LCDs. The DEFINE statements I used here should work with most LCDs, since I really slowed things down.

Next we establish the variables "Row," "Location," "Value," and "Char" which will be used throughout the PBPro code. After that, the Init label sets up PORTB for proper data direction using the TRIS directive. This is followed by a direct control of PORTB to set the state of each PORTB pin. Finally, we use the LCDOUT command to display an initial message that says "Serial LCD".

The label `start` is where this project code really starts. The main line is the `SERIN` line that waits for the row, location, and value bytes. The program will sit here forever if it doesn't receive any information. When information is received it first tests the "Row" byte and "Location" byte to see if "Row" is not 0 and "Location" is 0 by using an `IF-THEN` statement. If these variables are at that state, it indicates don't move the cursor from its existing position and write the character in the "Value" byte at that location. It does that by jumping to the `Display` label.

If the "Location" byte is not 0, independent of the "Row" byte value, then the next command run is a `BRANCH` instruction. The `BRANCH` instruction jumps the program to the proper label based on the value of the "Row" byte. If the "Row" byte is 0, then it jumps to the `Command` label. The code at the `Command` label will send a LCD command based on the "Value" byte received.

If "Row" does not equal 0, then the `BRANCH` command redirects the program to the `Row1` or `Row2` labels. At label `Row1` and `Row2`, the program converts the "Location" byte received into the proper LCD code to position the LCD cursor at the proper row and column using the `LOOKUP` command. We have to subtract 1 from the "Location" byte because the `LOOKUP` command starts at 0 instead of 1.

After we have the proper position code from the `LOOKUP` command, the program then jumps to the `LCDcmd` subroutine to send the special LCD position code to the LCD and move the cursor on the LCD. When the subroutine is done, the program jumps back to the command after the `GOSUB LCDcmd` line. That command is a jump to the `Display` label.

At the `Display` label, the program first stores the serially received "Value" byte and stores a copy of it into the "Char" variable. Then the program jumps to the subroutine `WrLCD` to send that byte to the LCD character generator that actually displays the character on the LCD. The program then returns to the routine at the `Display` label, which sends the program back to the top to receive a new set of information at the `Start` label. Note how we first positioned the cursor based on the "Row" and "Location" bytes and then sent the character code ("Value" byte) to be displayed at that position.

As mentioned above, if the "Row" byte is zero, the `BRANCH` command under the `Start` label redirects the program to the `Command` label. We don't stop at the `Row1` and `Row2` labels to get a LCD command code because the "Value" byte received serially should have the LCD command byte in it. All we have to do is convert that "Value" byte to the proper LCD command code based on the table above the Command label. We take the "Value" byte and convert it into a decimal number by subtracting hex $30. Then we use the LOOKUP command to change the "Value" byte into the proper LCD command code. That command code is stored in the "Char" variable.

The next command line jumps the program to the LCDcmd subroutine where the command code is sent to the LCD. After that is complete, we return from the subroutine and then jump back to the Start label to receive more data.

```
' —-[ Title ]——————————————————--
'
' File...... proj9pro.BAS
' Purpose... Serial -> PIC16F876 -> LCD (4-bit interface)
' Author.... Chuck Hellebuyck
' Started... January 22 2002
' Updated...


' —-[ Program Description ]——————————————-
'
'
' PIC16F876 Port Hardware connections:
'
' PIC            LCD                  Other Connections
' ———           ———-                 ————————————--
' B4          LCD.11
' B5          LCD.12
' B6          LCD.13
' B7          LCD.14
' B3          LCD.4
' B0          LCD.6
' OSC1                            Resonator - 4 mhz
' OSC2                            Resonator - 4 Mhz
' MCLR                 Vdd via 1k resistor
' Vdd                          5v
' Vss                          Gnd
' B1                                      Max232(RX)
' B2                                      Max232(TX)
```

```
' —-[ Revision History ]————————————
'
'
'—-[DEFINES]—————————————————

include "modedefs.bas"      'include serout defines
Define LOADER_USED 1        'Only required if bootloader used to
                            'program PIC

DEFINE LCD_DREG PORTB       'Define PIC port used for LCD Data
                            'lines
DEFINE LCD_DBIT     4            'Define first pin of portb
                                 'connected to LCD DB4
DEFINE LCD_RSREG PORTB      'Define PIC port used for RS line of
                            'LCD
DEFINE LCD_RSBIT 3          'Define Portb pin used for RS
                            'connection
DEFINE LCD_EREG     PORTB   'Define PIC prot used for E line of LCD
DEFINE LCD_EBIT     0            'Define PortB pin used for E
                                 'connection
DEFINE LCD_BITS     4            'Define the 4 bit communication
                                 'mode to LCD
DEFINE LCD_LINES 2          'Define using a 2 line LCD
DEFINE LCD_COMMANDUS 2000   'Define delay between sending LCD
                            'commands
DEFINE LCD_DATAUS 50             'Define delay time between data
                                 'sent.

' —-[ Constants ]————————————————--
'
' —-[ Variables ]————————————————--
'
ROW var byte          ' LCD ROW value
LOCATION var byte     ' Column position on the LCD
VALUE var byte        ' Value is the ASCII Character to display
char var byte              ' Temporary storage of character code

' —-[ Initialization ]————————————
'
Init:
      TRISB = $0000                    ' all outputs off to start
      portb = %11111101            ' LCD pins
      LCDOUT "Serial LCD"          ' Display project name on LCD
      pause 1000                         ' Delay 1 second
```

```
' —-[ Main Code ]————————————————--
'
'Display initial screen
'
'******* Main Program ***************************************

Start:
        SERIN 1,T2400,ROW,LOCATION,VALUE 'Receive serial data
            Row = Row - $30                   'Correct Row to
                                              'decimal number
            Location = Location - $30      'Correct Location to
                                           'decimal number


'******* Decision and Branch Routine **************************
' if ROW = 0 then its a command so jump to command
' if ROW does not = 0 and LOCATION = 0 then write the value where
' the cursor is. VALUE is the ASCII equivalent of the value sent
' per the LCD ASCII chart

IF ROW <> 0 AND LOCATION = 0 THEN DISPLAY        'Test for zero
                                                 'value
BRANCH ROW,[COMMAND,Row1,Row2]',Row3,Row4] 'Branch to proper row

'******* position cursor where new value will be written *********

Row1:
        LOCATION = LOCATION - 1     'Correct location for zero value
'**** Shrunk line to fit, convert location byte to char byte for
                                    'LCD command
LOOKUP LOCATION, [$80,$81,$82,$83,$84,$85,$86,$87,$88,$89,$8A,$8B,
    $8C,$8D,$8E,$8F,$90,$91,$92,$93],char
        GOSUB LCDcmd                'Jump to LCD command routine
        GOTO DISPLAY                ' Jump to LCD display routine

Row2:
        LOCATION = LOCATION - 1     'Correct location for zero value
'**** Shrunk line to fit, convert location byte to char byte for
LCD command
LOOKUP LOCATION, [$C0,$C1,$C2,$C3,$C4,$C5,$C6,$C7,$C8,$C9,$CA,$CB,
    $CC,$CD,$CE,$CF,$D0,$D1,$D2,$D3],char
        GOSUB LCDcmd                'Jump to LCD command routine
        GOTO DISPLAY                'Jump to LCD command routine
'
'**** These commented out lines are for converting this program to
                    ' 4x16 LCDs
```

```
'Row3:
'          LOCATION = LOCATION - 1
'LOOKUP LOCATION,[$94,$95,$96,$97,$98,$99,$9A,$9B,$9C,$9D,$9E,$9F,
'$A0,$A1,$A2,$A3,$A4,$A5,$A6,$A7],char
'          GOSUB LCDcmd
'          GOTO DISPLAY
'
'Row4:
'          LOCATION = LOCATION - 1
'LOOKUP LOCATION,[$D4,$D5,$D6,$D7,$D8,$D9,$DA,$DB,$DC,$DD,$DE,$DF,
'$E0,$E1,$E2,$E3,$E4,$E5,$E6,$E7],char
'          GOSUB LCDcmd
'          GOTO DISPLAY


'****** convert value to be displayed for wrlcd routine
****************

DISPLAY:
          char = VALUE      'Store Value byte into char variable
          GOSUB WrLCD       'Jump to routine that sends display
                            'characters to LCD
          GOTO START        'Jump back to beginning of main loop

'***** ROW=0, Therefore run a command per list below
*******************

'0      clear LCD and move to position 1 of row 1
'1      shift cursor left
'2      Display is off, Cursor is off, Cursor Blink is off
'3      Display is on, Cursor is off, Cursor Blink is off
'4      Display is on, Cursor is on, Cursor Blink is off
'5      Display is on, Cursor is off, Cursor Blink is on
'6      Display is on, Cursor is on, Cursor Blink is on
'7      shift display right
'8      shift display left
'9      shift cursor right

COMMAND:
          value = value - $30       'Convert value to decimal number for
                                    'command only
'**** Convert Value byte to LCD code byte and store in char ***
          LOOKUP VALUE,[$01,$10,$08,$0C,$0E,$0D,$0F,$1C,$18,$14],char
          GOSUB LCDcmd 'Jump to routine that sends LCD commands
          GOTO START   'Jump to beginning of program


'  *** Send command byte to LCD Subroutine
```

```
'
LCDcmd:
        LCDOUT $FE, char                ' Send command to LCD
        RETURN                  'Return to where this routine was
                                'called


' *** Send ASCII character, to be displayed, to LCD
'
WrLCD:
        LCDOUT char             ' Send char to LCD
        RETURN                  ' Return to where this routine was
                                ' called


        END
```

Final Thoughts

I have to admit this isn't the best serial LCD example I could have written, but it is unique in that each position of the LCD can be accessed with a single serial command based on a simple row and column (location) value. This project can be expanded in numerous ways. Just playing around with the way the program receives data could be a major change. Possibly making the SERIN command line part of a loop until a certain character is received would allow multiple characters to be received at one time before putting them on the LCD screen. Use your imagination!

Hopefully you've learned how to control a LCD module and how to communicate using the SERIN and SEROUT commands. These are some of the most common functions your PicBasic programs will perform and also brings PICs into the real world. People ask me all the time what I do with the programming tools I use and sell at my website. Once I show them a PC screen responding to actions at the PIC circuit, it seems to answer their question. When the LCD displays messages to them, they seem to understand this as well. Soon they are making comments about building alarm systems and sprinkler systems, or even ideas for robots. It's the human connection they were looking for and the LCD or PC is just that. Adding an LCD and serial connection to any PIC project makes it look more professional and a lot more complicated than it actually is. But PicBasic makes it easy!

CHAPTER **8**

Memory and Sound

In this chapter we explore both the internal memory of the PIC and access to external memory. The memory I refer to is neither the program memory (where your code is stored) nor the data or RAM memory (where your variables are accessed). I'm referring to the EEPROM (*Electrically Erasable Programmable Read Only Memory*). This is the memory that stores information or data you want to keep alive after the power is disconnected. It can be changed at any time within a program, but once the PIC is shut down anything in EEPROM is not erased.

All flash memory PICs have internal EEPROM memory, but usually less than 1k. If you need more than 1k, then external EEPROM memory chips are used. You can get large 128k memory chips in a tiny 8-pin package.

Access to the internal EEPROM is quite easy in PicBasic by using a single WRITE or READ command. External memory is a bit tougher, though. Many external memory chips communicate using the Phillips I²C protocol. PicBasic makes that easy with the I2CIN and I2COUT commands.

This chapter's projects will show how to access the internal EEPROM memory and access external memory. After that, we'll do something completely different. We make the PIC into a music generator by playing music through a speaker using the SOUND command.

Project #10—Using External Memory

In addition to PIC microcontrollers, Microchip also makes external EEPROM chips. The most common types of these chips are controlled using the I²C protocol, and the PicBasic I2CIN and I2COUT commands make communicating with these chips quite easy.

This project is easy to build, but is quite handy when combined with other functional code. Just as previous projects did, this project forms the basis for using a PIC to access external keep-alive memory. The program simply accesses external memory and sends it out serially to a terminal program running on a PC.

We use a Microchip 24LC00 device in this project, which is actually a very small memory part. It only has 16 bytes of EEPROM space. You might ask why we use such a small chip when the internal PIC memory is larger than that. The reason is to show how simple EEPROM can be added to any PIC, especially PICs that don't have any internal EEPROM. Even 16 bytes of EEPROM can be very handy for very little cost. The 24LC00 works the same as other EEPROM chips, so the code is the same. Besides, I had a free 24LC00 sample from Microchip lying around! Figure 8-1 shows the schematic diagram for this project and Figure 8-2 shows the completed circuit board.

Hardware

The PBC and PBPro manuals show the connections required to use the I2COUT and I2CIN commands. This project uses those connections to the external Microchip 24LC00 I²C memory chip. RA0 and RA1 must be used for the PBC program because those connections are predefined in the PBC internal code. The group of connections to ground on the EEPROM chip's A0, A1, and A2 pins are required to designate the address of the chip.

The schematic in Figure 8-1 also shows the serial connection using the RB2 and RB1 pins. Those can be moved to any pins you want but the code below would have to change. They feed into a RS232 level shifter chip like the one used in Chapter 7 projects. This chip converts the 0–5 volt PIC signals into +12 to −12 volt signals the PC can easily read. The last two connections are the SCL and SDA pins. These are the serial clock and serial data lines. They are the communication lines used to send data to the EEPROM chip.

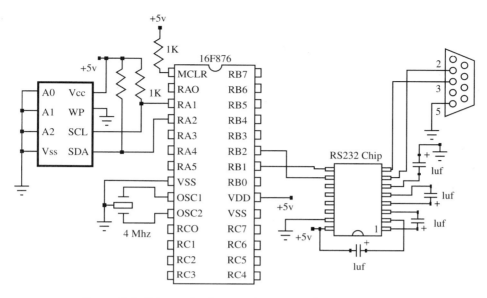

Figure 8-1: Schematic diagram for external memory project.

Figure 8-2: View of the completed circuit shown in Figure 8-1.

PBC Code

The program first initializes symbols by setting up the address and value bytes. The serial in and serial out connections are made into constants "SI" and "SO". This makes it easy to change the serial connections later by just changing these two constants to match the hardware.

Next the EEPROM control byte is established. This can be found in the PBC manual or in the EEPROM data sheet. If just one of these bits is off, the whole program can act screwy or not work at all. Following that the "address", "value", and temporary "x" variables are established. As a final step, we establish the constant for the ADCON1 register in the PIC memory map. We'll use that to set up PORTA.

Now the program does one more step before entering the `init` label. We issue the command line below:

```
poke adcon1, 7
```

This line is required because the PIC initializes PORTA to analog mode. In other words, the port is automatically set up to act as an A/D port. By modifying the ADCON1 register with a %00000111or 7 using the `POKE` command, we set PORTA to all digital mode.

(This has everything to do with the PIC and nothing to do with PBC.)

Now the program enters the `init` label. At this label we have a `FOR-NEXT` loop that initializes all 16 bytes of EEPROM data to a value of 10. We simply increment the address variable and the use the `I2COUT` command to send the value in the parenthesis, which is 10, to each value of address in the EEPROM.

The main loop is entered at the `Rx` label. At this label, the program first sends out a line of instructions for the command format using the `SEROUT` command. The format is the address (0–15) followed by the value to be stored (0–254). The program then waits for two bytes of data that can be sent by any serial communication device in an 8N1 format (8 data bits, no parity, 1 stop bit) at 2400 baud. A PC running a terminal program works great for this.

Some terminal programs will allow you to send numbers larger than 9 with a "#" symbol in front of it. This will send the actual number rather than the ASCII value of each character. For example, if you send 10 then the ASCII value $31 followed by $30 would be sent, which is the ASCII value for "1" and then "0." If instead #10 is sent, then the value 10 or $0A is sent. The latter is what you want to make this program work. Check your terminal program to see how to send the data.

The program only accepts one byte of data at a time. Using the I2COUT command, the program then sends the address and data bytes to the 24LC00 chip. The 24LC00 finds that address and then stores the data there. If the address byte received ever equals 255, then the program ignores the data sent and jumps to the Tx label.

At the Tx label, the external memory is read starting at location 0 using the I2CIN command. The data is retrieved byte by byte and as each byte is received, the SEROUT command is used to send each byte out to the terminal program. It continues until all 16 locations have been read and sent. Then the program jumps back to the Rx label to wait for new data.

As I noted at the beginning of this section, this is a simple program but can be very useful.

```
' —-[ Title ]————————————————————-
'
' File...... Proj10PB.BAS
' Purpose... PIC16F876 -> 24LC00 Microchip EEPROM
' Author.... Chuck Hellebuyck
' Started... February 9, 2002
' Updated...


' —-[ Program Description ]————————————
'
' This program demonstrates the use of the I2CIN and I2COUT
' commands.
'
```

```
' The program is written to work with a PC. The PC will send
' serial data to the circuit to be stored into the 16 bit EEPROM
' chip. If a value of ' 255 is sent to the circuit, then the
' program will read each EEPROM location and send the data to the
' PC to be displayed. The receiving part of the program requires
' the address and the value of the data to be stored. It then
' overwrites that address with the received data using the I2COUT
' command. When the data is 255, the program uses the I2CIN to
' read the EEPROM one byte at a time and send that data to the PC.
'
' Hardware Connections:
'
' PIC            EEPROM Pin      EEPROM Pin Name      Misc Conn.
' ___            _____ _____  _____
' RA0            EEPROM.5        SDA
' RA1            EEPROM.6        SCL
' RB1                                         Serial In (RX)
' RB2                                         Serial Out (TX)
'                EEPROM .1       A0            Gnd
'                EEPROM. 2       A1            Gnd
'                EEPROM. 3       A2            Gnd
'                EEPROM. 4       Vss           Gnd
'                EEPROM. 7       WP            Gnd
'                EEPROM. 8       Vcc           5v
'
' ---[ Revision History ]--------------------------
'
'
'
' ---[ Defines ]--------------------------------
'
'
'
' ---[ Variables and Constants ]------------------
'
symbol SO = 2                   ' Define serial output pin
symbol SI = 1                   ' Define serial input pin
symbol control = %01010000      ' Set EEPROM control byte
symbol Address = b3             ' Byte to store address
symbol Value = b2               ' Byte to store value to store
symbol X = b1                   ' multi-purpose variable
symbol adcon1 = $9f             ' Define adcon1 register address
'
```

```
' —-[ Initialization ]————————————————

        poke adcon1, 7                  ' Set Port A to digital
                                        ' I/O
Init:
For address = 0 To 15           ' Loop 16 times
        I2Cout control,address, (10)  ' Preset each address to ten
        Pause 10                      ' Delay 10ms after each write
        Next


' —-[ Main Code ]————————————————


RX:
serout SO, T2400,("Enter #address#value") 'Display instruction
                                          'line
Serin SI,T2400,address, value           ' Receive location and
                                        ' data to store
If address = 255 then TX        ' Test for data dump request
I2Cout control, address, (value)  ' Store value received at
address pause 10                ' Delay to allow write to occur

Goto RX                                 ' Jump back to receive more
                                        ' data




TX:
For address = 0 To 15                   ' Loop thru 16 locations
I2Cin control, address, value           ' Read byte at address X
Serout SO,T2400,(#address, ": ",#value, 10, 13)
Next                                    ' Send Address and
                                        ' value received
                                        ' read to PC

Goto RX                                 ' Loop back to the top to
                                        ' receive
                                        ' data
```

PBCPro Code

Before we initialize anything in the PBCPro version of the code, including variables
and constants, we have to issue a special `include` statement:

```
Include "modedefs.bas"          ' Include serial modes
```

Unlike PBC, PBPro does not predefine the baud rate mode names such as T2400 or N2400. I find them to be really handy and easy to understand later when I'm reviewing code. PBPro allows you to use them if you add that include line.

After that we add the usual DEFINE for PBPro to recognize we are using a boot-loader:

```
define loader_used 1            'Used for bootloader only
```

The program first initializes symbols by setting up the address and value bytes. The serial in and serial out connections are made into constants "SI" and "SO". This makes it easy to change the serial connections later by just changing these two constants to match the hardware.

Next the EEPROM control byte is established. This can be found in the PBPro manual or in the EEPROM data sheet. If just one of these bits is off, the whole program can act screwy or not work at all.

We add the alias names DPIN and CPIN to Port A pin 0 and 1 which are our data and clock pins. This makes reading the I2CIN and I2COUT commands easier to follow later. Following that the "address", "value", and temporary "x" variables are established.

Prior to entering the init label, we issue the command lines below. Because PBPro recognizes PIC register names, we don't have to define what address the ADCON1 register is at in PIC memory. We just act on the ADCON1 register directly to make Port A digital pins. After that we have to establish Port A direction registers as outputs.

```
adcon1 = 7              ' Set PortA to digital ports
        TRISA = %00000000   ' Set PortA as all outputs
```

Now the program enters the init label. At this label we have a FOR-NEXT loop that initializes all 16 bytes of EEPROM data to a value of 20. We simply increment the address variable and then use the I2COUT command to send the value in the brackets, which is 20, to each value of address in the EEPROM.

The main loop is entered at the Rx label. At this label the program first sends out a line of instructions for the command format using the SEROUT command. The format is address (0-15) followed by the value to be stored (0-254). The program then waits for two bytes of data that can be sent by any serial communication device in an 8N1 format (8 data bits, no parity, 1 stop bit) at 2400 baud. A PC running a terminal program works great for this.

Some terminal programs will allow you to send numbers larger than 9 with a "#" symbol in front of it. This will send the actual number rather than the ASCII value of each character. For example, if you send 10, then the ASCII value of $31 followed by $30 would be sent, which is the ASCII value for "1" and then "0". If #10 is sent instead, then the value 10, or $0A is sent. This is how you want this program work. Check your terminal program to see how to send the data.

The program only accepts one byte of data at a time. Using the I2COUT command, the program then sends the address and data bytes to the 24LC00 chip. The 24LC00 finds that address and then stores the data there. If the address byte received ever equals 255, then the program ignores the data sent and jumps to the Tx label.

At the Tx label, the external memory is read starting at location 0 using the I2CIN command. The data is retrieved byte by byte and, as each byte is received, the SEROUT command is used to send each byte out to the terminal program. It continues until all 16 locations have been read and sent. Then the program jumps back to the Rx label to wait for new data. As was true with the PBC version, this is a simple but useful program.

```
' —-[ Title ]——————————————————————————————--
'
' File...... Proj10PR.BAS
' Purpose... PIC16F876 -> 24LC00 Microchip EEPROM
' Author.... Chuck Hellebuyck
' Started... February 9, 2002
' Updated...


' —-[ Program Description ]————————————————————
'
' This program demonstrates the use of the I2CIN and I2COUT
' commands.
'
```

```
' The program is written to work with a PC. The PC will send
' serial data to the circuit to be stored into the 16 bit EEPROM
' chip. If a value of 255 is sent to the circuit, then the program
' will read each EEPROM location and send the data to the PC to be
' displayed. The receiving part of the program requires the
' address and the value of the data to be stored. It then
' overwrites that address with the received data using the I2COUT
' command. When the data is 255, the program uses the I2CIN to
' read the EEPROM one byte at a time and send that data to the PC.
'
' Hardware Connections:
'
'
' PIC           EEPROM Pin      EEPROM Pin Name     Misc Conn.
' ___           _____ _____                     _____
' RA0           EEPROM.5        SDA
' RA1           EEPROM.6        SCL
' RB1                                               Serial In (RX)
' RB2                                               Serial Out (TX)
'               EEPROM .1       A0                  Gnd
'               EEPROM. 2       A1                  Gnd
'               EEPROM. 3       A2                  Gnd
'               EEPROM. 4       Vss                 Gnd
'               EEPROM. 7       WP                  Gnd
'               EEPROM. 8       Vcc                 5v

' ---[ Revision History ]----------------------------
'
'


' ---[ Defines ]------------------------------
'
        Include "modedefs.bas"        ' Include serial modes
        define loader_used 1          'Used for bootloader only


' ---[ Variables and Constants ]------------------
'
SO      con     2                     ' Define serial output pin
SI      con     1                     ' Define serial input pin
Control con     %10100000             ' Set EEPROM control byte
```

```
DPIN      var      PORTA.0              ' I2C data pin
CPIN      var      PORTA.1              ' I2C clock pin
Address   var      byte                 ' Byte to store address
Value     var      byte                 ' Byte to store value to
store
X         var      byte                 ' multi-purpose variable

' ---[ Initialization ]-----------------------------
'

    adcon1 = 7           ' Set PortA to digital ports
    TRISA = %00000000    ' Set PortA as all outputs

Init:
For x = 0 To 15                         ' Loop 16 times
        I2Cwrite dpin,cpin,control,x,[20]    ' Preset each
                                        ' address to ten
        Pause 10                        ' Delay 10ms after
                                        ' each write

        Next

' ---[ Main Code ]----------------------------------


RX:
serout SO, T2400,["Enter #address#value"] 'Display instruction
                                        ' line
Serin SI,T2400,address, value           ' Receive location
                                        ' and data to
                                        ' store
If address = 255 then TX                ' Test for data dump
                                        ' request
I2Cwrite dpin,cpin,control,address,[value]' Store value received
                                        ' at address received

pause 10                                ' Delay to allow
                                        ' write to
                                        ' occur

Goto RX                                 ' Jump back to
                                        ' receive more
                                        ' data
```

```
TX:
For X = 0 To 15                          ' Loop thru 16 locations
I2Cread dpin,cpin,control, X, [value]    ' Read byte at address X
Serout SO,T2400,[#X, ": ",#value, 10, 13] ' Send Address and value
                                         ' read to PC

Next

Goto RX                                            ' Loop back to the
                                                   ' top
                                                   ' to receive data
```

Final Thoughts

This project can easily be used as the basis for other projects that need configuration data stored for later retrieval when the PIC is first powered up. RAM data, such as variables, need to be cleared or initialized at the beginning of a program. EEPROM data can be used to preload those variables with data developed when the PIC was last running.

You can also use one of the larger memory chips if necessary, but check the PicBasic manual since the control byte in the I2CIN and I2COUT commands will need to be matched to the chip you use.

Project #11—Accessing Internal Memory

This project functions similarly to the previous project, except we use the internal EEPROM space of the 16F876 PIC. However, this doesn't work for all PICs; it only works with the PICs that have internal EEPROM, which are typically the flash memory PICs. The project uses the READ and WRITE commands. Because the memory is internal to the PIC, we don't have to worry about external connections or control bytes. All we need is the address byte and the data byte.

As Microchip develops new PICs, the size of the internal EEPROM available seems to grow. If you only need 256 bytes or less, you can probably find a PIC with enough internal EEPROM memory. Figure 8-3 shows the schematic and Figure 8-4 illustrates the completed circuit.

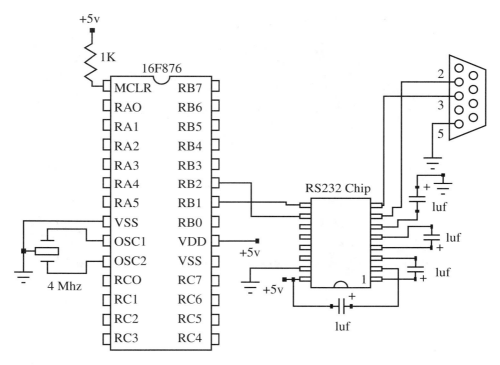

Figure 8-3: Schematic diagram for accessing internal memory.

Hardware

The hardware is similar to the previous project. We use the RB2 and RB1 pins for SEROUT and SERIN communication by tying them to a RS232 level shifter. We also have the standard power, resonator, and MCLR pull-up connections. Other than that, however, the rest of the connections are internal to the PIC.

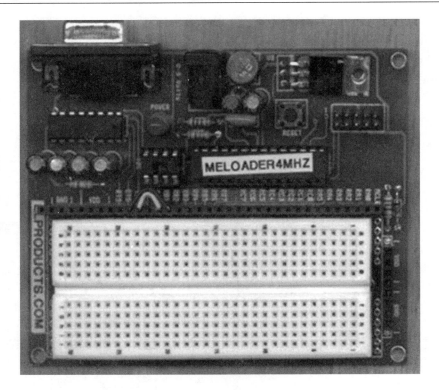

Figure 8-4: View of the completed circuit shown in Figure 8-3.

PBC Code

To program this circuit using the PBC compiler, we first initialize the constants and variables that make it easy to follow. SO and SI are defined, as the SEROUT and SERIN pins. After that we reserve space and define the "Address" variable and the "Value" variable.

The next step is the Init label section of code. We use a FOR-NEXT loop with a WRITE command in the middle of the loop to initialize the first 16 bytes of internal EEPROM to all 10's. This is similar to the previous Project #10, but now we use the WRITE command.

When the initialization of the memory is finished, the program jumps to the Rx label. At that label is a SERIN command waiting for address byte and data byte from a source communicating at 2400 baud 8N1 format. A terminal program does the job here again.

When the address byte is received, the program tests the address byte value to see if it is equal to 255. If it is equal to 255, then the program jumps to the TX label and begins to dump the contents of internal EEPROM using the READ command. Each address is read within a FOR_NEXT loop. The data byte read is then sent serially in a 2400-baud 8N1 format that can be displayed with a terminal program. When all 16 address bytes have been read and sent, the program jumps back to the RX label to get more incoming data.

If the address received in the RX label's FOR-NEXT loop is not 255, then the program proceeds to store the "Value" byte received at the "Address" byte received using the WRITE command. The program then loops back to receive another set of data.

```
' —-[ Title ]————————————————————————————————-
'
' File...... Proj11PB.BAS
' Purpose... PIC16F876 -> Internal PIC16f876 EEPROM
' Author.... Chuck Hellebuyck
' Started... February 9, 2002
' Updated...


' —-[ Program Description ]————————————————————-
'
' This program demonstrates the use of the Read and Write
' commands. The program is written to work with a PC. The PC will
' send serial data to the circuit to be stored into the PICs
' internal EEPROM. If a value of 255 is sent to the circuit, then
' the program will read each EEPROM ' location and send the data
' to the PC to be displayed.
'
' To update the existing EEPROM with new data, the receiving part
' of the program requires the address and the value of the data to
' be stored. It then overwrites that address with the received
' data using the Write command. When the data is 255, the program
' uses the Read command to read the EEPROM one byte at a time and
' send that data to the PC.
```

```
'
' Hardware Connections:
'
' Since everything is internal to the PIC there is no external
' hardware to hookup. Just the standard Oscillator, MCLR, power
' and ground connections.
'
' ---[ Revision History ]-----------------------------------
'
'
'
' ---[ Defines ]--------------------------------------------'
' ---[ Variables and Constants ]---------------------------
'
symbol SO = 2                          ' Define serial output pin
symbol SI = 1                          ' Define serial input pin
symbol Address = b3                    ' Byte to store address
symbol Value = b2                      ' Byte to store value to store
' ---[ Initialization ]-------------------------------------
'
Init:
For address = 0 To 15                  ' Loop 16 times
      Write address, 10                  ' Preset each address to ten
      Pause 10                           ' Delay 10ms after each write
      Next

' ---[ Main Code ]------------------------------------------
'


RX:
serout SO, T2400,("Enter #address#value") ' Display instruction
line
Serin SI,T2400,address, value                    ' Receive location
                                                 ' and data
                                                 ' to store

If address = 255 then TX                   ' Test for data dump
                                           ' request

Write address, value                            ' Store value
                                                ' received at
                                                ' address received

pause 10                                        ' Delay to allow
                                                ' write to
                                                ' occur
```

```
Goto RX                                      ' Jump back to receive
                                             ' more
                                             ' data


  TX:
For address = 0 To 15              ' Loop thru 16 locations
       Read address, value              ' Read byte at address X
       Serout SO,T2400,(#address, ": ",#value, 10, 13) ' Send
Address and
                                      ' value read to PC

       Next

       Goto RX                               ' Loop back to the
                                             ' top to
                                             ' receive data
```

PBPro

Before we initialize anything, including variables and constants, we have to issue a special `include`:

```
Include "modedefs.bas"           ' Include serial modes
```

Unlike PBC, PBPro does not predefine the baud rate mode names such as T2400 or N2400. However, I find them to be really handy and easy to understand later when I'm reviewing code. PBPro allows you to use them if you add the `include` line above.

Before we enter the main program section, we initialize the constants and variables that make it easy to follow. `SO` and `SI` are defined, as the `SEROUT` and `SERIN` pins. After that, we reserve space and define the "Address" variable and the "Value" variable. The next step is the `Init` label section of code. We use a `FOR-NEXT` loop with a `WRITE` command in the middle of the loop, to initialize the first 16 bytes of internal EEPROM to all 10's. This is similar to Project #10, but now we use the `WRITE` command.

Once the initialization of the memory is finished, the program jumps to the Rx label. At that label is a SERIN command waiting for the address byte and data byte from a source communicating at 2400 baud 8N1 format. A terminal program does the job here again.

When the address byte is received, the program tests the address byte value to see if it's equal to 255. If it's equal to 255, then the program jumps to the TX label and begins to dump the contents of internal EEPROM using the READ command. Each address is read within a FOR_NEXT loop. The data byte read is then sent serially in a 2400 baud 8N1 format that can be displayed with a terminal program. When all 16 address bytes are read and sent, the program jumps back to the RX label to get more incoming data.

If the address received in the RX label's FOR-NEXT loop is not 255, then the program stores the "Value" byte received at the "Address" byte received using the WRITE command. The program then loops back to receive another set of data.

```
' —-[ Title ]————————————————————--
'
' File...... Proj10PR.BAS
' Purpose... PIC16F876 -> Internal 16F876 EEPROM
' Author.... Chuck Hellebuyck
' Started... February 9, 2002
' Updated...


' —-[ Program Description ]————————————
'
' This program demonstrates the use of the Read and Write
' commands. The program is written to work with a PC. The PC will
' send serial data to the circuit to be stored into the PICs
' internal EEPROM. If a value of 255 is sent to the circuit, then
' the program will read each EEPROM location and send the data to
' the PC to be displayed. To update the existing EEPROM wih new
' data, the receiving part of the program requires the address and
' the value of the data to be stored. It then overwrites that
' address with the received data using the Write command. When the
' data is 255, the program uses the Read command to read the
' EEPROM one byte at a time and send that data to the PC.
```

```
'
' Hardware Connections:
'
' Since everything is internal to the PIC there is no external
' hardware to hookup. Just the standard Oscillator, MCLR, power
' and ground connections.
'
' ---[ Revision History ]----------------------------------
'
'
' ---[ Defines ]---------------------------------

        Include "modedefs.bas"          ' Include serial
                                        ' modes
        define loader_used 1           'Used for
                                        ' bootloader only


' ---[ Variables and Constants ]------------------------
'
SO       con     2                     ' Define serial output pin
SI       con     1                     ' Define serial input pin
Address  var     byte                  ' Byte to store address
Value    var     byte                  ' Byte to store value to
store

' ---[ Initialization ]----------------------------------
'

Init:
        For address = 0 To 15          ' Loop 16 times
        Write address,10          ' Preset each address to ten
        Pause 10                       ' Delay 10ms after each
                                       ' write

        Next

' ---[ Main Code ]--------------------------------------
'

RX:
serout SO, T2400,["Enter #address#value"] ' Display instruction
                                          ' line
Serin SI,T2400,address, value             ' Receive location
                                          ' and data to
                                          ' store
```

```
If address = 255 then TX          ' Test for data dump
                                  ' request
Write address,value               ' Store value received at
                                  ' address received
pause 10                                  ' Delay to allow
                                          ' write to
                                          ' occur

Goto RX                           ' Jump back to receive
                                  ' more
                                  ' data



    TX:
        For address = 0 To 15         ' Loop thru 16 locations
        Read address, value           ' Read byte at address X
        Serout SO,T2400,[#address, ": ",#value, 10, 13]      ' Send
Address and
                                      ' value read to PC

        Next

        Goto RX                           ' Loop back to the
                                          ' top to
                                          ' receive data
```

This program only uses the first 16 bytes of 16F876 internal EEPROM memory. It has 256 bytes so you can easily expand the FOR-NEXT loops to use all the memory if you want. Once again, this program can be used as the basis for other programs that may need configuration memory storage or data storage that has to stay available after power is removed. EEPROM memory is great for storing data you want to keep if the battery goes dead in your battery-powered PIC project.

The WRITE and READ commands are very similar between PBPro and PBC, so expanding one can easily be ported over as an improvement for the other.

Project #12—Making Music

"Making music," as the heading above says, might be a bit of a stretch. However, we will play "Mary Had a Little Lamb" as a series of computer beeps. The point of this project is to demonstrate how to use the SOUND command to produce audible sound through a speaker. It will sound similar to one of those greeting cards that play a tune when you open them. The tune is played through a series of codes retrieved by a LOOKUP command. Each of those codes is converted to a signal by the SOUND command to play a note through an 8-ohm speaker.

Hardware

Because the PIC has such high-power output capability, it can drive a speaker directly. The circuit adds a capacitor in series to block any DC and pass the AC tune through to the speaker. The hardware is fairly easy to build. Figure 8-5 shows the schematic diagram for this project and the completed circuit is shown in Figure 8-6.

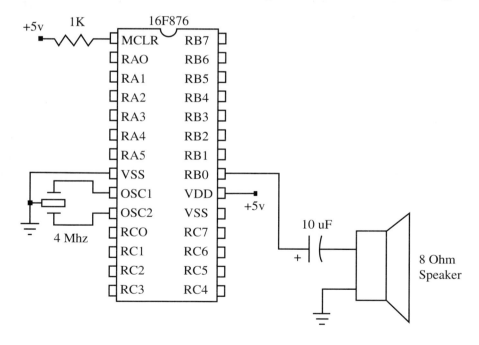

Figure 8-5: Schematic diagram for generating music with a PIC.

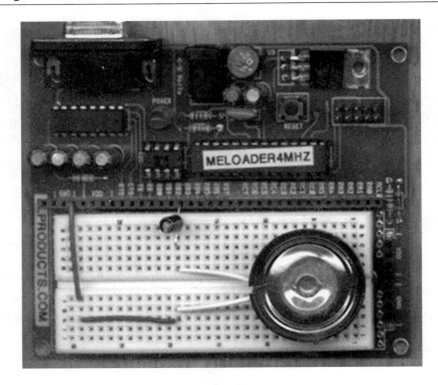

Figure 8-6: The completed music generation circuit.

As with previous PIC projects, we have the same MCLR pull-up resistor and resonator. We add the 8-ohm speaker, with a 10 μf capacitor in series, to Port B pin 0. The PicBasic manual shows this same circuit, and is recommended when using the SOUND command. Make sure you connect the capacitor with the correct polarity. Any 8-ohm speaker should work, but a small unit will probably be best.

PBC Code

The program starts by establishing the pin used for the SOUND command. It's the Port B pin connected to the speaker or Port B pin 0. By making it a constant with the SYMBOL directive, we can easily move the speaker on the hardware with very little revisions needed to the software.

Next we establish three variables "x", "tone", and "dur" which represent a counter variable to keep track of the number of notes, the tone variable (or note to be played), and the duration (or how long we play it). After that we DELAY for 1 second. There is no reason for doing this other than I wanted to have a pause before it started blasting out a tune!

The main loop is at the beep label. Here is where the SOUND command gets used. What we do is use a FOR_NEXT loop to step through 51 notes that are stored in the LOOKUP command's line. Notice how the LOOKUP command spreads across three lines here in the code listing that follows? These should be one continuous line in your actual program.

The FOR-NEXT loop first retrieves the note to be played in the first LOOKUP command line and then the duration is retrieved in the second LOOKUP command line. After that, the two codes are used by the SOUND command to play that note for that duration. The loop then jumps back up to retrieve another set of note and duration.

Once all the notes have been played, you should have heard a crude version of "Mary Had A Little Lamb." The program will wait 10 seconds for you to digest this wonderful symphony of sound and then it jumps back up to the top to do it all again.

```
' ---[ Title ]--------------------------------------
'
' File...... Proj12PB.BAS
' Purpose... PIC16F876 -> Speaker
' Author.... Chuck Hellebuyck
' Started... February 9, 2002
' Updated...


' ---[ Program Description ]------------------------
'
' This program demonstrates the use of the Sound Command.
'
' The program is written to play a crude version of "Mary had a
' little lamb". You will find the PIC is not the greatest musician
' but this program demonstrates how the sound command can add
' noise or audio feedback to a button press or an error in data
' entry.
```

```
'
' PIC                 Misc Conn.
' ___                 _____
' RB0                 + side of 10uf cap
'                     - side of 10uf cap to speaker
'                     other speaker wire to ground
'
' ---[ Revision History ]--------------------
'
'
'
' ---[ Defines ]-----------------------------
'
'
'
' ---[ Variables and Constants ]-------------
'
Symbol   SND = 0            ' 10uf cap to Speaker Pin
Symbol   tone = b3          ' Variable for storing tone
Symbol   dur = b5           ' Variable for storing duration
Symbol   x = b1             ' General counting variable

' ---[ Initialization ]----------------------
'
Init:
    Pause 1000              ' Let PIC delay before blasting its
                            ' tune

' ---[ Main Code ]---------------------------
'


Beep:
    for x = 0 to 50  ' Step thru 51 total tones

'** The for next loop steps thru 51 notes and durations using two
'** lookup commands. The lines are so long, I had to add comments
'here.
'** I also had to break them up into three lines. Your program
'must
'** make this one continuous long line or it won't work.

lookup x,(80,0,75,0,68,0,75,0,80,0,80,0,80,0,75,0,
75,0,75,0,80,0,80,0,80,0,80,0,75,0,68,0,75,0,80,0,
80,0,80,0,80,0,75,0,75,0,80,0,75,0,68),tone

lookup x,(80,0,80,0,80,0,80,0,80,0,80,0,80,0,80,0,
```

```
80,0,80,0,80,0,80,0,80,0,80,0,80,0,80,0,80,0,80,0,
80,0,80,0,80,0,80,0,80,0,80,0,80,0,80),dur

Sound SND,(tone,dur)          ' Generate Sound
next                          ' Get next note and duration

pause 10000                   ' Delay for 10 seconds for applause
                              ' before playing again

Goto beep                     ' Do it forever
```

PBPro

The PBPro version of the program starts off by establishing the pin used for the SOUND command. It's the Port B pin connected to the speaker or Port B pin 0. By making it a constant with the SYMBOL directive, we can easily move the speaker or the hardware with very little revisions to the software.

Next we establish three variables "x", "tone", and "dur", which represent a counter variable to keep track of the number of notes, the tone variable (or note to be played), and the duration (or how long we play it). After that we DELAY for 1 second because I wanted to have a pause before it started blasting out a tune.

The main loop is at the beep label. Here is where the SOUND command gets used. What we do is use a FOR_NEXT loop to step through 51 notes that are stored in the LOOKUP command's line. Notice how the LOOKUP commands spreads across three lines in the code listing that follows? Unlike PBC above, PBPro allows this if you enter an underscore ("_") character after each broken line. This allows long command lines to fit in a small format that is easy to read and print.

The FOR-NEXT loop first retrieves the note to be played in the first LOOKUP command line. Then the duration is retrieved in the second LOOKUP command line. After that, the two codes are used by the SOUND command to play that note for that duration. The loop then jumps back up to retrieve another set of note and duration.

Once all the notes have been played, you should have heard a crude version of "Mary Had A Little Lamb." The program will wait 10 seconds for you to enjoy the magnificent sound and then it will jump back up to the top to do it all again.

This program is not much different than the previous PBC version. The biggest differences are the variable set-up, the LOOKUP command, and SOUND command. Remember that LOOKUP and SOUND use brackets instead of parentheses in PBPro.

```
' —-[ Title ]————————————————-
'
' File...... Proj12PR.BAS
' Purpose... PIC16F876 -> Speaker
' Author.... Chuck Hellebuyck
' Started... February 9, 2002
' Updated...


' —-[ Program Description ]————————-
'
' This program demonstrates the use of the Sound Command.
'
' The program is written to play a crude version of "Mary had a
' little lamb". You will find the PIC is not the greatest musician
' but this program demonstrates how the sound command can add
' noise or audio feedback to a button press or an error in data
' entry.
'
' PIC         Misc Conn.
' ___         _____-
' RB0           + side of 10uf cap
'               - side of 10uf cap to speaker
'               other speaker wire to ground
'
' —-[ Revision History ]——————————
'
'
' —-[ Defines ]————————————-
'
      define loader_used 1              'Used for bootloader
                                        'only

' —-[ Variables and Constants ]———————————-
'
SND con 0          ' 10uf cap to Speaker Pin
x var byte         ' Temporary counter byte
tone var byte        ' Variable for storing tone
dur var byte         ' Variable for storing duration
```

```
' ---[ Initialization ]--------------------------
'
Init:
    Pause 1000              ' Let PIC delay before blasting its
                            ' tune
' ---[ Main Code ]------------------------------- --
'


Beep:
    for x = 0 to 50  ' Step thru 51 total tones

'****** The for next loop steps thru 51 notes and durations using
'two lookup
'****** commands. The lines are so long, I had to add comments
'here.
    lookup x,[80,0,75,0,68,0,75,0,80,0,80,0,80,0,75,0,75,_
    0,75,0,80,0,80,0,80,0,80,0,75,0,68,0,75,0,80,0,80,0,80,_
    0,80,0,75,0,75,0,80,0,75,0,68],tone

    lookup x,[80,0,80,0,80,0,80,0,80,0,80,0,80,0,80,0,80,_
    0,80,0,80,0,80,0,80,0,80,0,80,0,80,0,80,0,80,0,80,0,80,_
    0,80,0,80,0,80,0,80,0,80,0,80],dur

    Sound SND,[tone,dur]    ' Generate Sound
    next                    ' Get next note and duration
    pause 10000             ' Delay for 10 seconds for applause
before playing again
    Goto beep      ' Do it forever
```

Final Thoughts

You can simply change the existing tune by changing the values in the LOOKUP table. To make it easier to read, you could store the song code in external EEPROM memory and then access the song like we accessed data from external EEPROM in Project #10. One way to do this would be to have several different EEPROM chips, each with a different tune stored in them. To change the tune, just change the EEPROM chip.

By now you should be getting very comfortable writing PBC and/or PBPro programs. If you still are a bit shaky, then try playing around with the code given so far in this book to modify it using your own techniques. You'll learn more by doing that than any book can teach you. You'll make mistakes, but by catching those mistakes you will learn to program PICs in PicBasic better than you can by just reading about PicBasic.

CHAPTER 9

Robotics

Robots are not my area of expertise, but they have become very popular among electronics hobbyists in recent years. The inspiration for this chapter was a recent trip to an assembly plant where I saw a three-wheeled robot in action, as shown in Figure 9-1.

Figure 9-1: An industrial robot in action.

The robot had two rear drive wheels and a heavy-duty caster wheel in front. The robot pulled a series of trailers that carried parts from one end of the plant to the other. It followed a line that was buried into the concrete floor. It also had an obstacle detection system in front that was simply a clear plastic shield with switches attached via cables. If anything bumped into the plastic, the cables would release tension to the switches and signal the robot to stop.

I'd seen hobbyist robot kits like this before, but never really took them too seriously. After I got home, though, I immediately starting putting together the pieces to build my own PicBasic controlled version. It's shown in Figure 9-2.

Figure 9-2: A PicBasic-controlled robot.

In this chapter, we will discuss three robotics projects. The first will simply control a three-wheeled robot platform with two rear drive wheels and one front caster wheel. The second project will add a line-following sensor to determine if a white or black color is present in front of the sensor; the purpose will be to build a basic robot that can follow a black line on the ground. The third project will add obstacle detection, but of greater complexity than the bumper switch method. It will use a Sharp GP2D15 infrared detector mounted on top of a servomotor. The servo will be mounted in the front of the robot and will sweep the detector back and forth to look for obstacles.

These will be three interesting projects (both to programmers and nonprogrammers), so let's get started!

Project #13—Robot Base

I'm not going to go through all the steps needed to build this robot platform because it's very easy to do (and this is a book about PicBasic, not robot construction!). There are several good books about robot construction, and much information about the subject is available on the World Wide Web. If you've never experimented with robots before, those are great sources for finding the information you'll need.

The hardest part of constructing the hardware for these projects will be reworking the servomotors, that is, if you do it yourself. As we discussed in a previous chapter, servomotors are designed to move back and forth based on a pulse-width modulated signal. In order to use these motors for a robot drive, their internals have to be reworked to spin a full 360 degrees in both directions. The robot tires are then attached to the shaft of the servomotors, so by individually turning the servomotors we can make the robot go forward, backward, turn right, and turn left. The rest of these robot projects can be built from parts available at a hardware store or hobby shop (a trip to your local electronics store goes without saying). Check out Appendix A for more on-line sources of robotics parts and complete kits.

Reworked servomotors accept the same signal as a standard servomotor; the key difference is that when the signal pulse is 1.5 milliseconds long, the servomotor stops turning. If the signal is greater than 1.5 milliseconds to 2.0 milliseconds, the motor turns counter-clockwise continuously. If the signal is less than 1.5 milliseconds to 1.0 milliseconds, then the servomotor will turn clockwise continuously.

As noted before, the robot has tires attached to the shaft of the servomotors and by individually turning the servomotors we can make the robot go forward, backward, turn right, and turn left. That is what this project is all about. We will use the PicBasic PULSOUT command to send the control signals to the servomotors and make the robot follow the following preset pattern:

1) First, the robot will drive forward for a short distance.

2) Then the robot will turn left about 180 degrees.

3) Next, the robot will drive backwards a short distance.

4) Finally it will turn right about 180 degrees.

5) The program will then loop back up to the top and do it all over again.

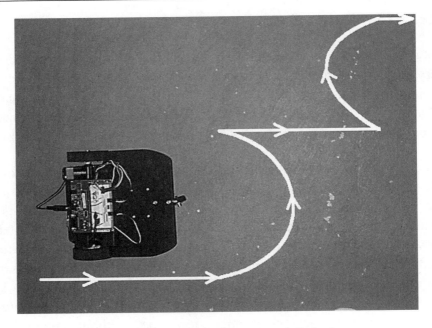

Figure 9-3: Path of the PicBasic-controlled robot.

Figure 9-3 shows the robot platform and the direction it will follow. The schematic in Figure 9-4 shows the electrical connections to the PIC16F876. Once again the standard connections are the same, with MCLR tied to Vdd through a resistor and the Vdd and ground connections to the power source. The servomotors draw a lot of power, so I put them and the PIC circuit on separate battery supplies. The ground points are connected, however, so they all share the same reference point.

Because servomotors have internal circuitry that handles the high current of the motors, a PIC pin can directly communicate with the servo. PIC pin B2 controls the right wheel servo while B7 controls the left wheel. These simple connections allow us to drive the robot.

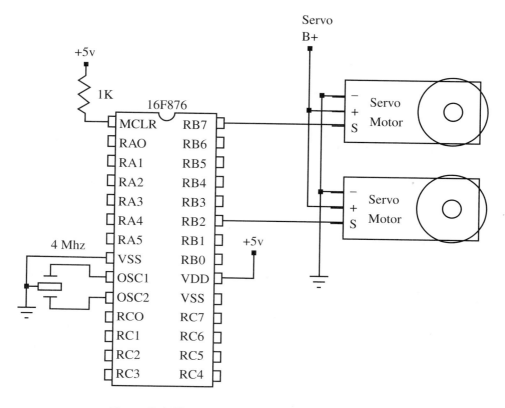

Figure 9-4: Electrical connections to the PIC16F876.

PBC Code

The program is really quite simple once you understand the key strategies to driving the servomotors. Because the servomotors are on opposite sides of each other you have to spin the left wheel counter-clockwise and the right wheel clockwise to make the robot go straight. You reverse that when you drive the robot in reverse. By just spinning one wheel at a time, you make the robot turn.

The PULSOUT command is used to send either 200 for counterclockwise rotation or 100 for clockwise rotation; the code creates four different subroutines for handling direction of the robot. This allows a simple main loop that calls each of the subroutines as the desired direction is needed. The four subroutines are labeled FWD, REVS, RFWD, and LFWD. They stand for *forward direction, reverse direction, turn right in forward direction*, and *turn left in forward direction*. These form the "control system" for the robot.

The program first establishes two byte-size variables: "move" and "x". These will be used throughout the program. Then PortB is set up with the DIRS directive to make all PortB pins outputs. After that the main program loop is entered.

The first block of commands is similar to the rest, so I'll explain the first block below:

```
' *** Move robot forward and pause
move = 100        'Preset move to 100 for distance
gosub fwd         'Jump to fwd subroutine for forward movement
pause 10          'Delay before next move
```

The variable "move" is given a value of 100. This will actually be used by the servomotor control subroutine as a means to move the robot wheels. I did not calculate how much each value would equate to in distance the robot will travel; I just tried different "move" values until I got what I wanted. A large value for "move" will drive the robot wheel further, while a small value will turn the wheel a shorter distance.

After the move value is established, the program jumps to the FWD subroutine. After the program returns from the subroutine, it pauses for a short time before the next directional movement is sent. The FWD subroutine is representative of all the subroutines, so let's examine that next:

```
' *** Move robot forward subroutine
FWD:
For x = 1 to move      'Start forward movement loop
pulsout 2, 100         'Turn right wheel forward
pulsout 7, 200         'Turn left wheel forward
pause 10               'Delay to control robot speed
next                   'Repeat forward movement loop
return                 'Jump back to where this subroutine was called
```

At the FWD subroutine, the "move" variable is used as the end point of a FOR-NEXT loop. The "move" variable simply controls how many times the servomotor is turned on by this subroutine, and thus controls the distance the robot wheel travels.

Within the FOR-NEXT loop are the PULSOUT commands that turn the robot wheels. They just send a pulse to the servomotors that their internal circuitry reads and then uses to drive the servomotor shaft. Since the servomotors are reworked, the position that matches the pulse width value sent is never seen by the servomotors' internal circuitry. That makes the wheel continue to turn and drive the robot wheel every time it gets a signal.

After the PULSOUT commands have pulsed the robot wheels a short distance, the subroutine has a PAUSE command. This pause slows down the loop time and thereby controls the robot speed. Make this value too small and you won't slow it down much at all; make it too large and the robot will be jumpy. This was strictly a trial and error effort to arrive at a delay of 10 msec.

When the FOR-NEXT loop is complete, the subroutine uses the RETURN command to jump back to the main loop. The program returns at the command following the GOSUB command that jumped the program to the subroutine.

The main loop continues this flow for the right turn, reverse and left turn routines that make up this project. When all those routines are complete, the main loop starts all over again at the top with a goto main command line. The robot will perform that same pattern all the way across the floor until you pick it up and pull the power.

```
' —-[ Title ]————————————————————————————-
'
' File...... proj13pb.BAS
' Purpose... Drive Robot in Unique Pattern PIC16F876 -> Servos
' Author.... Chuck Hellebuyck
' Started... March 1, 2002
' Updated...


' —-[ Program Description ]————————————————-
'
' This Program uses the 16F876 to control a Servo robot platform.
' The robot has dual servo motors with wheels mounted to drive the
' robot. This program will drive the robot forward a short
' distance and then turn left. After completing the left turn, the
```

```
' robot will drive straight in reverse for a short distance and
' then stop. After that the robot will turn right. After
' completing the right turn, the robot will start the routine over
' with the forward movement.
'
' RB2          Right Wheel Servo
' RB7          Left Wheel Servo

' ---[ Revision History ]----------------------------------
'
'

' ---[ Constants ]----------------------------------------
'

' ---[ Variables ]----------------------------------------
'
symbol move = b0                'robot movement variable
symbol x = b1                   'general purpose variable

' ---[ Initialization ]------------------------------
'

Init:
DIRS = %1111111111              'Set portB to all outputs

' ---[ Main Code ]----------------------------------------
'

main:
' *** Move robot forward and pause
move = 100      'Preset move to 100 for distance
gosub fwd       'Jump to fwd subroutine for forward movement
pause 10        'Delay before next move

' *** Make robot turn left 180 degrees and pause
move = 250      'Preset move to 250 for 180 degrees turn
gosub lfwd      'Jump to lfwd subroutine to turn left
pause 10        'Delay before next move

' *** Move robot in reverse and then pause
move = 100      'Preset move to 100 for distance backwards
gosub revrs     'Jump to revrs subroutine to move backwards
pause 10        'Delay before next move
```

```
' *** Make robot turn right 180 degrees and pause
move = 200          'Preset move to 200 for 180 degrees turn
gosub rfwd          'Jump to rfwd subroutine to turn right
pause 10            'Delay before next move

goto main           'Loop back and do it again


' *** Move robot forward subroutine
FWD:
For x = 1 to move   'Start forward movement loop
pulsout 2, 100      'Turn right wheel forward
pulsout 7, 200      'Turn left wheel forward
pause 10            'Delay to control robot speed
next                'Repeat forward movement loop
return              'Jump back to where this subroutine was called

' *** Move robot in reverse subroutine
revrs:
For x = 1 to move   'Start reverse movement loop
pulsout 2, 200      'Turn right wheel in reverse
pulsout 7, 100      'Turn left wheel in reverse
pause 10            'Delay to control robot speed
next                'Repeat reverse movement loop
return              'Jump back to where this subroutine was called

' *** Move robot forward-left subroutine
Lfwd:
For x = 1 to move   'Start left-forward movement loop
pulsout 2,100       'Turn right wheel forward to go left
pause 10            'Delay to control robot speed
next                'Repeat left-forward movement loop
return              'Jump back to where this subroutine was called

' *** Move robot forward-right subroutine
Rfwd:
For x = 1 to move   'Start right-forward movement loop
pulsout 7, 200      'Turn left wheel forward to go right
pause 10            'Delay to control robot speed
next                'Repeat right-forward movement loop
return              'Jump back to where this subroutine was called
```

PBPro Code

The PBPro program version is very similar to the PBC program. There are areas that could have been simplified, but I wanted to try to maintain a common structure so the PBPro version copies the PBC version.

As with the PBC code, the program is really quite simple once you understand the key strategies for driving the servomotors. Because the servomotors are on opposite sides of each other, you have to spin the left wheel counter-clockwise and the right wheel clockwise to make the robot go straight. You reverse that when you drive the robot in reverse; by just spinning one wheel at a time, you make the robot turn.

The PULSOUT command is used to send either 200 for counter-clockwise rotation or 100 for clockwise rotation. What the code does is create four different subroutines for handling direction of the robot. This allows a simple main loop that calls each of the subroutines according to which direction is needed. The four subroutines are labeled FWD, REVS, RFWD, and LFWD. They stand for *forward direction, reverse direction, turn right in forward direction,* and *turn left in forward direction.* These form the main control of the robot.

Starting at the top, the program establishes two byte-size variables: "move" and "x". These will be used throughout the program. Then PortB is set with the TRISB directive to make all PortB pins outputs. After that, the define loader_used 1 line is issued to let the compiler know we are using a bootloader to program the PIC. Then the main loop of code is entered.

The first block of commands is similar to the rest, so I'll explain the first block below:

```
' *** Move robot forward and pause
move = 100      'Preset move to 100 for distance
gosub fwd       'Jump to fwd subroutine for forward movement
pause 10        'Delay before next move
```

The variable "move" is given a value of 100. This will actually be used by the servomotor control subroutine as a means to move the robot wheels. I did not calculate how much each value would equate to in distance the robot will travel. A large value for "move" will drive the robot wheel further, while a small value will turn the wheel a shorter distance.

After the "move" value is established, the program jumps to the FWD subroutine. After the program returns from the subroutine, the program pauses for a short time before the next directional movement is sent.

The FWD subroutine represents all the subroutines, so we'll examine that next:

```
' *** Move robot forward subroutine
FWD:
For x = 1 to move      'Start forward movement loop
pulsout 2, 100         'Turn right wheel forward
pulsout 7, 200         'Turn left wheel forward
pause 10               'Delay to control robot speed
next                   'Repeat forward movement loop
return                 'Jump back to where this subroutine was called
```

At the FWD subroutine, the "move" variable is used as the end point of a FOR-NEXT loop. The "move" variable simply controls how many times the servomotor is turned on by this subroutine and thus controls the distance the robot wheel travels.

Within the FOR-NEXT loop are the PULSOUT commands that turn the robot wheels. They send a pulse to the servomotors that their internal circuitry reads and uses to drive the motor shaft. Since the servomotors are reworked, the position that matches the pulse width value sent is never seen by the servomotors' internal circuitry. That makes the wheel continue to turn and drive the robot wheel every time it gets a signal.

After the PULSOUT commands have pulsed the robot wheels a short distance, the subroutine has a PAUSE command. This slows down the loop time and thus controls the robot speed. Make this value too small and you won't slow it down much at all; make it too large and the robot will be jumpy. This was strictly a trial and error effort to arrive at a delay of 10 msec.

When the FOR-NEXT loop is complete, the subroutine uses the RETURN command to jump back to the main loop. The program returns at the command following the GOSUB command that jumped the program to the subroutine.

The main loop continues this flow for the right turn, reverse, and left turn routines that make up this project. When all those routines are complete, the main loop starts all over again at the top with a GOTO main command line. The robot will perform that same pattern all the way across the floor until you pick it up and pull the power.

```
' ----[ Title ]------------------------------
'
' File...... proj13pr.BAS
' Purpose... Drive Robot in Unique Pattern PIC16F876 -> Servos
' Author.... Chuck Hellebuyck
' Started... March 1, 2002
' Updated...


' ----[ Program Description ]----------------
'
' This Program uses the 16F876 to control a Servo robot platform.
' The robot has dual servo motors with wheels mounted to drive the
' robot. This program will drive the robot forward a short
' distance and then turn left. After completing the left turn, the
' robot will drive straight in reverse for a short distance and
' then stop. After that the robot will turn right. After
' completing the right turn, the robot will start the routine over
' with the forward movement.
'
' RB2       Right Wheel Servo
' RB7       Left Wheel Servo

' ----[ Revision History ]-------------------
'
'


' ----[ Constants ]--------------------------
'

' ----[ Variables ]--------------------------
'
move var byte               'robot movement variable
x var byte                  'general purpose variable

' ----[ Initialization ]---------------------
```

```
'

Init:
TRISB = %00000000          'Set portB to all outputs

' ——-[ Main Code ]————————————————————--
'
        define loader_used 1              'Used for bootloader only

main:
' *** Move robot forward and pause
move = 100        'Preset move to 100 for distance
gosub fwd         'Jump to fwd subroutine for forward movement
pause 10          'Delay before next move

' *** Make robot turn left 180 degrees and pause
move = 250        'Preset move to 250 for 180 degrees turn
gosub lfwd        'Jump to lfwd subroutine to turn left
pause 10          'Delay before next move

' *** Move robot in reverse and then pause
move = 100        'Preset move to 100 for distance backwards
gosub revrs       'Jump to revrs subroutine to move backwards
pause 10          'Delay before next move

' *** Make robot turn right 180 degrees and pause
move = 200        'Preset move to 200 for 180 degrees turn
gosub rfwd        'Jump to rfwd subroutine to turn right
pause 10          'Delay before next move

goto main         'Loop back and do it again


' *** Move robot forward subroutine
FWD:
For x = 1 to move    'Start forward movement loop
pulsout 2, 100       'Turn right wheel forward
pulsout 7, 200       'Turn left wheel forward
pause 10             'Delay to control robot speed
next                 'Repeat forward movement loop
return               'Jump back to where this subroutine was called

' *** Move robot in reverse subroutine
revrs:
For x = 1 to move    'Start reverse movement loop
pulsout 2, 200       'Turn right wheel in reverse
```

```
pulsout 7, 100      'Turn left wheel in reverse
pause 10            'Delay to control robot speed
next               'Repeat reverse movement loop
return             'Jump back to where this subroutine was called

' *** Move robot forward-left subroutine
Lfwd:
For x = 1 to move   'Start left-forward movement loop
pulsout 2,100      'Turn right wheel forward to go left
pause 10           'Delay to control robot speed
next               'Repeat left-forward movement loop
return             'Jump back to where this subroutine was called

' *** Move robot forward-right subroutine
Rfwd:
For x = 1 to move   'Start right-forward movement loop
pulsout 7, 200     'Turn left wheel forward to go right
pause 10           'Delay to control robot speed
next               'Repeat right-forward movement loop
return             'Jump back to where this subroutine was called
```

Final Thoughts

The projects that follow are really follow-on steps to this project. Before continuing further, I suggest you play with the `main` loop and make the robot drive different patterns. Changing the "move" variable values and the pause values in the subroutine is something else to try. From that, you will have a better understanding of the effects these values have on your robot platform. (I mention this because not all servomotors will react the same way.)

Project #14—Line Tracker

This project expands the previous project's robot platform, so I'll omit details of the robot construction. When I was taking pictures for the hardware section below, I tried out a really lousy digital camera. The picture was fuzzy and not usable for what I intended, but something interesting was captured with that camera. It had caught the infrared light shining down from the infrared detectors. For the description to follow, this picture is really worth a thousand words. See it in Figure 9-5.

Figure 9-5: The infrared detectors are how the robot tracks the "test track."

I found this project to be quite a lot of fun, but also a challenge. I found a line-follower kit that was fairly easy to assemble and easy to interface to a PIC. It had three infrared emitter/detector pairs separated on the board about 0.75-inch apart. It also has buffer electronics that output three signals, with one signal output pin for each detector. When the detectors are placed about 0.25-inch above a white surface, the output is high (5 volts). When the detector is placed above a black surface, the output is low (ground). By monitoring the three outputs, we should be able to control the robot motors to keep the center detector above a black line about 0.75 inch in width. That is exactly what this project does.

To start, I drew a long curvy line on a string of printer paper taped together. This formed the "test track" on top of my bench for running the line follower. The first time I tested my code, it worked quite well. But after playing with it a while, I found some bugs. The hardest part was debugging when the robot would suddenly stop in the middle of the test track. Because of that, I added three LEDs to signal which detector pair was over the black line.

When the center detector was over the black line, I lit a green LED, when the left detector was over the black line, I lit a yellow LED, and when the right detector was over the black line I lit a red LED. Every once in a while, the robot would stop, yet the LEDs showed the green LED lit. This implied to me that the robot should keep moving forward, as the center detector was over the black line as desired. Unfortunately, that wasn't the case.

I discovered the robot was following the code I wrote rather than the code I thought I wrote! It turned out I hadn't written code to handle the situation of when none of the detectors sensed the black line, which is what was happening when the black line was thin enough to fit between two of the sensors. The green LED was lit from the previous loop of code, not because the sensor was over the black line. To correct this problem, I could have simply made the line wider. Instead, I modified the code to move forward until the line was found again. This also allowed the robot to run around freely on a white surface until a black line was found. It seems simple now, but it took me more than an hour to figure out my mistake. Once that was corrected, the robot followed the line perfectly.

The LEDs also added an unplanned visual "special effect." As the robot drove across my bench test track, the LEDs would scramble back and forth, showing the robot going too far to the left or right and then correcting itself back to the center, as the robot zigzagged down the line. I showed my two young sons and their friends the robot with the idea they might ask how it worked (I also wanted to demonstrate what dad had been working on in the basement). Instead they were more fascinated by the LEDs that scrolled back and forth quickly! They seemed to understand the line following much better than I had expected and without any coaching from me. (I probably should have just showed them the scrolling LED project.)

Neither the PBC nor the PBPro code is super complex; it just expands on previous chapter examples of reading digital I/O. It does, however, offer a real-world example of interfacing to a sensor, which is really at the heart of robotic control. Figure 9-6 shows the schematic diagram while Figure 9-7 shows the robot tracking the path thanks to the completed circuit.

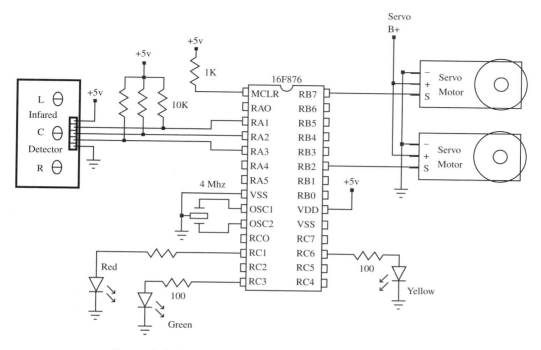

Figure 9-6: Schematic diagram of the line tracking circuit.

Figure 9-7: The completed line tracking circuit in action.

Hardware

This circuit expands on the one for the previous project to add circuitry to follow the line. I show it as a block diagram since the infrared circuitry is contained in a kit. It has three infrared detector pairs soldered to the board. Each one has an infrared LED and an infrared detector transistor built together inside of a small package. The infrared light from the LED shines down and is reflected back off of the test track surface. The infrared detector transistor receives the reflected light and turns on if the light is strong enough. Since black absorbs light and white reflects light, the black line will not reflect any light, so the detector transistor will not turn on. The transistor will turn on when the detector is over the white portion of the test track.

The biggest hang-up is getting the proper height set for the detector. If the sensor is too far away from the track, the reflection will not be strong enough and inconsistent signals will result. If it's too close, the reflection will not be picked up. Once you get it set, it will work great. The added circuitry in the detector module cleans up the transistor signal and outputs the low or high signal the PIC monitors. For example, when the detector is over black and the transistor is off, the circuitry drives the output low for the PIC to read.

The detector module has three outputs, one for each infrared detector. The PIC uses Port A to read those three outputs. The left detector output is tied to the A1 pin. The center detector output is tied to the A2 pin. The right detector is tied to the A3 pin. The module also needs power and ground. Because the detector module draws a small amount of current, the module gets powered off the same 5-volt source used to power the PIC.

The red, yellow, and green LEDs used for monitoring the detectors are tied to the PortC pins. The right detector is monitored with a yellow LED connected to pin C6. The center detector is monitored with a green LED connected to pin C3. The left detector is monitored with a red LED and connected to C1.

Port B controls the servomotors in the exact same way Project #13 did. They are once again on a separate power supply than the PIC but have the grounds connected.

PBC Code

This program actually worked the first time I tried it. Then I attempted to improve it and messed the whole thing up! I fortunately kept the original file and went back to it when all else failed. Through that experience, I found several bugs that the first trial run missed. As I mentioned earlier, the bugs would only show up when the sensor was in a position where all of the sensors were sensing white. It pays to save the original file.

Let's examine this program as chunks of code, similar to the way I described the previous project. First the constants are established:

```
symbol LSENS = bit1        'Left detector element
symbol RSENS = bit2        'Right detector element
symbol CSENS = bit3        'Center detector element
symbol adcon1 = $9F        'A/D control register location
symbol trisa = $85         'PortA Tris register location
symbol trisc = $87         'PortC Tris register location
symbol porta = 5           'PortA register location
symbol portc = 7           'PortC register location
```

LSENS, RSENS, and CSENS are the line detector sensor inputs. I tied them to bits 1, 2, and 3 within the predefined PBC byte B0. ADCON1 is the PIC Port A control register that we use to set Port A as digital pins. PBC does not allow direct control of that register, so to make the code easier to understand the name ADCON1 is tied to memory location $9F where the ADCON1 register resides in the PIC.

Next we make the port control easier to understand by defining the locations for Port A and Port C. Because PBC only works directly on Port B, we setup the TrisA and TrisC locations along with the Port A and Port C data registers. We will later use POKE and PEEK to control these ports.

At the next chunk we establish the same variables used in Project #13, "move" and "x".

```
symbol move = b3           'robot movement variable
symbol x = b4              'general purpose variable
```

Now the program enters the initialization phase at the init label:

```
Init:
poke ADCON1, 7             'Set portA to digital I/O
poke TRISA, %11111111      'Set portA to all inputs
poke TRISC, %00000000      'Set portC to all outputs
DIRS = %11111111           'Set portB to all outputs
```

The program uses POKE to set up the ports. The ADCON1 register is set to %00000111 or decimal 7. This makes Port A into digital pins. Then the program "POKEs" TrisA with all ones to make Port A all inputs. Port A is now ready to run.

Port C is next. The program "POKEs" TrisC with all zeros to make all of Port C into outputs. Finally the DIRS directive is used to make all of Port B into outputs. Note that Port B is the only port that uses 1 as output set-up and 0 as input set-up; this was explained in earlier chapters but I wanted to remind you of this.

Now the main code loop is entered.

```
main:
peek porta, b0
if csens = 1 and lsens = 1 and rsens = 1 then error    'Can't find
                                                        'line
if lsens = 0 and rsens = 0 then halt          'Both elements
                                              'found line
if csens = 0 then center     'Center element on line goto center
                             'routine
if lsens = 0 then left          'Left element on line goto left
                                'routine
if rsens = 0 then right          'Right element on line goto
                                 'right routine
goto main                              'Loop back and do it
                                       'again
```

The first line uses the PEEK command to read Port A and store the result in predefined variable "B0". I could have given "B0" a better name using the symbol directive, but I wanted you to understand how reading that variable is used to read the line detector. Remember that Bits 1, 2, and 3 of the "B0" variable are assigned to the detector pins. This only works when B0 is filled from a Port A read (or PEEK).

Pin 1 of the detector is connected to Port A pin1. Therefore when Port A is "PEEKed" into B0, Bit 1 holds the state of Port A pin 1. By reading these bits, we can determine which detectors see black line and which see white space. That is exactly what the next set of commands does.

Through a series of IF-THEN commands, we test each of the inputs as a group and also individually. First the program checks if all inputs see white space ("1"). The first IF-THEN statement handles this by using the "AND" logical within the IF-THEN statement as seen below:

```
if csens = 1 and lsens = 1 and rsens = 1 then error    'Can't find
                                                       'line
```

If all of the detectors see a high level, then the program jumps to the following error routine:

```
' *** All elements not on line error routine
error:
poke portc, %01001010    'All LEDs on
move = 1                       'Preset move variable to 1
gosub fwd                      'Jump to fwd subroutine
goto main                      'Jump back to main loop
```

This error routine first sets all three LEDs to on, indicating the robot is in the error mode. That is done in one line by "POKEing" the Port C data register with the proper bits. I used the binary "%" directive so I could easily read which bits were 1 and which were 0.

Next the routine sets the "move" variable to one. This is a small number to make the robot creep forward. Then the routine jumps to the FWD subroutine. We do this so the robot will continue to slowly creep forward looking for the line. Since the robot never drives perfectly straight, the assumption is the robot will eventually find the line with one of the sensors. It worked quite well. The FWD subroutine is the same one discussed in the previous project:

```
FWD:
For x = 1 to move    'Start forward movement loop
pulsout 2, 100       'Turn right wheel forward
pulsout 7, 200       'Turn left wheel forward
pause 10             'Delay to control robot speed
next                 'Repeat forward movement loop
return               'Jump back to where this subroutine was called
```

At the forward subroutine, the "move" variable is used as the end point of a FOR-NEXT loop. Therefore the "move" variable just controls how many times the servomotor is turned on by this subroutine and thus controls the distance the robot wheels travel.

Within the FOR-NEXT loop are the PULSOUT commands that turn the robot wheels. They just send a pulse to the servomotors that their circuitry reads and then drives the motor shaft. Since the servomotors are reworked, the position that matches the pulse width value sent is never seen by their internal circuitry. That makes the wheel continue to turn and drive the robot wheel every time it gets a signal.

After the PULSOUT commands have pulsed the robot wheels a short distance, the subroutine has a PAUSE command. This pause slows down the loop time and thus controls the robot speed. Make this value too small and you won't slow it down much, but if you make it too large the robot operates in a "jumpy" fashion. I used strictly a trial-and-error method to arrive at a delay of 10 msec.

When the FOR-NEXT loop is complete, the subroutine uses the RETURN command to jump back to the error routine. At that point, the next command in the error routine jumps back to the main loop so the detectors can be tested again.

The test for all detectors at once is purposely put first in the list of IF-THEN statements. This is because the robot needs to find a line before any of the other IF-THEN statements mean anything. The robot will continue to creep forward until the line is found.

What would happen if the black line were never found? The robot would continue to creep all the way off the white surface to a nonreflective surface. At that point, I wanted the robot to stop. The second IF-THEN line takes care of that.

```
if lsens = 0 and rsens = 0 then halt          'Both elements
found line
```

I figured if the two outer sensors did not see white reflective surface, the robot must have gone off the test track. At that point I wanted it to stop. It also made it easy to create a stop point on the test track by ending the black line with a 90-degree line to form a "T." Once the robot got to the end of the track, both outer detectors saw the top bar of the "T" and the robot would stop.

The IF-THEN statement handles this by using another "AND" function within the IF-THEN command. This time we only look at the outer detectors. If both are 0, then they do not see white and the program jumps to the halt label, as seen below:

```
' *** Outer elements sense black line
halt:
poke portc, %01000010       'Right,Left LED on, Center LED off
pause 10                    'Add delay for 10 msec
goto main                   'Jump back to main loop
```

The `halt` routine first turns on the outer two LEDs with a `POKE` command to the Port C data register. Then it pauses for 10 msec. After that, the program jumps back to the `main` loop to test the detectors again. If the robot is still seeing darkness at the outer detectors, then the `halt` routine runs again. It's a delay routine that waits until the robot it picked up and placed back on the white test track.

The next series of `IF-THEN` statements in the `main` loop test each detector individually. If the sensors see a 0, then the black line is detected, so they jump to separate routines.

```
if csens = 0 then center      'Center element on line goto
                              'center routine
if lsens = 0 then left        'Left element on line goto left
                              'routine
if rsens = 0 then right       'Right element on line goto
                              'right routine
```

The "CSENS" detector is the one we want to stay on the black line. If a 0 is sensed in that line, the robot moves forward using the routine at the `center` label. It's similar to other robot movement routines, so I won't go into its details. It just inches the robot forward by driving both wheels and then jumps back to the `main` loop to test the detectors again.

`LSENS` and `RSENS` are used to correct when the robot is going off the black line. If the "LSENS" detector sees black, then the program jumps to the `left` label. At that label, the robot's right wheel is driven to turn the robot back towards the black line. If the "RSENS" detector sees black, then the program jumps to the `right` label. At that label, the robot's left wheel is driven to turn the robot back towards the black line.

These three `IF-THEN` statements do most of the control to make the robot zig-zag along the black line. After all the `IF-THEN` statements are complete, the program jumps back to the `main` label using a `GOTO` command and starts checking the `IF-THEN` statements all over again.

```
' ---[ Title ]--------------------------------------------
'
' File...... proj14pb.BAS
' Purpose... Robot to follow Black Line PIC16F876 -> IRD Line
' Detector
' Author.... Chuck Hellebuyck
```

```
' Started... March 1, 2002
' Updated...


' —-[ Program Description ]————————————--
'
' This Program uses the 16F876 to control a Servo robot platform.
' The robot has an infrared 3 element line detector module mounted
' under its platform. This program will drive the robot forward
' while it scans for the black line under the detector. If the
' black line hits one of the outer elements, the robot will
' correct back to center as it moves forward. The object is to
' have the robot follow the line from one end  to the other.
'
' RA1        Left Detector Element Output
' RA2        Center Detector Element Output
' RA3        Right Detector Element Output
' RC1        Left LED
' RC3        Center LED
' RC6        Right LED
' RB2        Right Wheel Servo
' RB7        Left Wheel Servo

' —-[ Revision History ]————————————
'
'

' —-[ Constants ]————————————--
'
symbol LSENS = bit1        'Left detector element
symbol RSENS = bit2        'Right detector element
symbol CSENS = bit3        'Center detector element
symbol adcon1 = $9F        'A/D control register location
symbol trisa = $85         'PortA Tris register location
symbol trisc = $87         'PortC Tris register location
symbol porta = 5           'PortA register location
symbol portc = 7           'PortC register location

' —-[ Variables ]————————————--
'
symbol move = b3                   'robot movement variable
symbol x = b4                      'general purpose variable
```

```
' —-[ Initialization ]————————————————————
'


Init:
poke ADCON1, 7                      'Set portA to digital I/O
poke TRISA, %11111111          'Set portA to all inputs
poke TRISC, %00000000          'Set portC to all outputs
DIRS = %11111111                   'Set portB to all outputs


' —-[ Main Code ]————————————————————-
'


main:
peek porta, b0
if csens = 1 and lsens = 1 and rsens = 1 then error    'Can't find
                                                       'line
if lsens = 0 and rsens = 0 then halt          'Both elements found
                                                 'line
if csens = 0 then center               'Center element on line goto
                                          'center routine
if lsens = 0 then left                 'Left element on line goto left
                                          'routine
if rsens = 0 then right                'Right element on line goto
                                          'right routine
goto main                              'Loop back and do it again

' *** Center element on line routine
center:
poke portc, 0            'All LEDs off
poke portc, 3            'Center LED on
move = 5                 'Preset move variable to 5
gosub fwd                'Jump to fwd subroutine
goto main                'Jump back to main loop

' *** Left element on line routine
left:
poke portc, 0            'All LEDs off
poke portc, 1            'Left LED on
move = 5                 'Preset move variable to 5
gosub lfwd               'Jump to lfwd subroutine
goto main                'Jump back to main loop

' *** Right element on line routine
right:
poke portc, 0            'All LEDs off
poke portc, 6            'Right LED on
```

```
move = 5                        'Preset move variable to 5
gosub rfwd                      'Jump to rfwd subroutine
goto main                       'Jump back to main loop


' *** All elements not on line error routine
error:
poke portc, %01001010           'All LEDs on
move = 1                        'Preset move variable to 1
gosub fwd                       'Jump to fwd subroutine
goto main                       'Jump back to main loop


' *** Outer elements sense black line
halt:
poke portc, %01000010           'Right,Left LED on, Center LED off
pause 10                        'Add delay for 10 msec
goto main                       'Jump back to main loop


' *** Move robot forward subroutine
FWD:
For x = 1 to move               'Start forward movement loop
pulsout 2, 100                  'Turn right wheel forward
pulsout 7, 200                  'Turn left wheel forward
pause 10                        'Delay to control robot speed
next                            'Repeat forward movement loop
return                          'Jump back to where this subroutine was
                                'called


' *** Move robot in reverse subroutine
revrs:
For x = 1 to move               'Start reverse movement loop
pulsout 2, 200                  'Turn right wheel in reverse
pulsout 7, 100                  'Turn left wheel in reverse
pause 10                        'Delay to control robot speed
next                            'Repeat reverse movement loop
return                          'Jump back to where this subroutine was
                                'called


' *** Move robot forward-left subroutine
Lfwd:
For x = 1 to move               'Start left-forward movement loop
pulsout 2,100                   'Turn right wheel forward to go left
pause 10                        'Delay to control robot speed
next                            'Repeat left-forward movement loop
return                          'Jump back to where this subroutine was
                                'called
```

```
' *** Move robot forward-right subroutine
Rfwd:
For x = 1 to move          'Start right-forward movement loop
pulsout 7, 200             'Turn left wheel forward to go right
pause 10                   'Delay to control robot speed
next                       'Repeat right-forward movement loop
return                     'Jump back to where this subroutine was
                           'called
```

PBPro Code

This PBCPro program is very similar to the PBC version just presented. The biggest difference is the direct control of the registers rather than using the PEEK and POKE commands. This is yet another example of a PBPro advantage over PBC.

Let's go through this as chunks of code, similar to the way I described the previous project. First the constants are established using the VAR directive.

```
LSENS var porta.1          'Left detector element
RSENS var porta.2          'Right detector element
CSENS var porta.3          'Center detector element
LLED var portc.1           'Left LED pin
CLED var portc.3           'Center LED pin
RLED var portc.6           'Right LED pin
```

LSENS, RSENS, and CSENS are the labels for the line detector sensor inputs connected to Port A. Using the VAR directive, these labels are tied to the specific Port A pins they connect to. The next set of labels, LLED, CLED, and RLED, are connected to the Port C pins that drive the LEDs. Labels like this make the program easier to read although "porta.1," etc. is fairly clear in what it represents.

At the next chunk we establish the same variables used in Project #13, "move" and "x".

```
move var byte              'robot movement variable
x var byte                 'general purpose variable
```

Now the program enters the initialization phase at the `init` label:

```
        define loader_used 1            'Used for bootloader only

Init:
ADCON1 = 7                      'Set portA to digital I/O
TRISA = %11111111               'Set portA to all inputs
TRISC = %00000000               'Set portC to all outputs
TRISB = %00000000               'Set portB to all outputs
```

The program first defines `loader_used` 1 because we are once again using a bootloader to program the PIC. Then the ADCON1 register is set to %00000111 or decimal 7. This sets Port A as digital pins. Following that, the port direction registers are set directly; Port A is set to all inputs and Ports B and C are made into all outputs.

From that section the `main` code loop is entered:

```
main:
if csens = 1 and lsens = 1 and rsens = 1 then error 'Can't find
                                                     'line
if lsens = 0 and rsens = 0 then halt     'Both elements found line
if csens = 0 then center     'Center element on line goto center
                             'routine
if lsens = 0 then left       'Left element on line goto left
                             'routine
if rsens = 0 then right      'Right element on line goto right
                             'routine
goto main                    'Loop back and do it again
```

Through a series of IF-THEN commands we test each of the inputs as a group and also individually. First the program checks to see if all inputs see white space ("1"). The first IF-THEN statement handles this by using the "AND" logical within the IF-THEN statement as seen below.

```
if csens = 1 and lsens = 1 and rsens = 1 then error    'Can't find
                                                       'line
```

If all of the detectors see a high level, then the program jumps to the `error` routine shown below.

```
' *** All elements not on line error routine
error:
high rled          'Right LED on
high lled          'Left LED on
high cled          'Center LED on
move = 1           'Preset move variable to 1
gosub fwd          'Jump to fwd subroutine
goto main          'Jump back to main loop
```

This `error` routine first sets all the LEDs to on. That indicates the robot is in the "error mode". I did it with a set of HIGH commands but could have done it similar to the PBC program in one line such as:

```
portc = %01001010
```

It was easier to remember the LED name rather than the port position for the LEDs. Because PBPro allows me to use the HIGH command directly on Port C, I did it the less efficient way for code but more efficient for my mind.

Next the routine sets the "move" variable to 1. This is a small number to make the robot creep forward. Then the routine jumps to the FWD subroutine. We do this so the robot will continue to slowly creep forward looking for the line. Since the robot never drives perfectly straight, the assumption is the robot will eventually find the line with one of the sensors. It worked quite well.

The FWD subroutine is the same one discussed in the previous project:

```
FWD:
For x = 1 to move    'Start forward movement loop
pulsout 2, 100       'Turn right wheel forward
pulsout 7, 200       'Turn left wheel forward
pause 10             'Delay to control robot speed
next                 'Repeat forward movement loop
return               'Jump back to where this subroutine was called
```

At the forward subroutine, the move variable is used as the end point of a FOR-NEXT loop. Therefore the "move" variable simply controls how many times the servomotor is turned on by this subroutine and thus controls the distance the robot wheels travel.

Within the FOR-NEXT loop are the PULSOUT commands that turn the robot's wheels. They just send a pulse to the servomotors that are read by their internal circuitry and then drive their motor shafts. Since the servomotors are reworked, the position that matches the pulse width value sent is never seen by their internal circuitry. That makes the wheel continue to turn and drive the robot wheel every time it gets a signal.

After the PULSOUT commands have pulsed the robot wheels a short distance, the subroutine has a PAUSE command. This pause slows down the loop time and thus controls the robot speed. As we noted with previous projects and the PBC version, the robot's movement will be erratic if the value is too large or too small. I used trial and error to determine that a delay of 10 msec worked best.

When the FOR-NEXT loop is complete, the subroutine uses the RETURN command to jump back to the error routine. At that point the next command in the error routine jumps back to the main loop so the detectors can be tested again.

The test for all detectors at once is purposely put first in the list of IF-THEN statements. This is because the robot needs to find a line before any of the other IF-THEN statements can be executed. As a result, the robot will continue to creep forward until the line is found.

Now it's certainly possible that the black line is never found. The robot would continue to creep all the way off the white surface to a nonreflective surface. At that point I wanted the robot to stop. The second IF-THEN line takes care of that:

```
if lsens = 0 and rsens = 0 then halt      'Both elements found
                                          'line
```

I figured if the two outer sensors did not see white reflective surface, then the robot must have gone off the test track. At that point I wanted it to stop. It also made it easy to create a stop point on the test track by ending the black line with a 90-degree line to form a "T." Once the robot got to the end of the track, both outer detectors saw the top bar of the "T" and the robot would stop.

The IF-THEN statement handles this by using another "AND" function within the IF-THEN command. This time we only look at the outer detectors. If both are 0, then they do not see white and the program jumps to the halt label seen below:

```
' *** Outer elements sense black line
halt:
high rled          'Right LED on
high lled          'Left LED on
low cled           'Center LED off
pause 10           'Add delay for 10 msec
goto main          'Jump back to main loop
```

The `halt` routine first turns on the outer two LEDs with `high` commands to the PortC data register. Then it pauses 10 msec. After that the program jumps back to the main loop to test the detectors again. If the robot is still seeing darkness at the outer detectors, then the `halt` routine runs again. It waits until the robot is picked up and placed back on the white test track.

The next series of `IF-THEN` statements in the `main` loop test each detector individually. If the sensors see a 0, then the black line is detected, so they jump to separate routines:

```
if csens = 0 then center       'Center element on line goto
                               'center routine
if lsens = 0 then left         'Left element on line goto left
                               'routine
if rsens = 0 then right        'Right element on line goto
                               'right routine
```

The "CSENS" detector is the one we want to stay on the black line. If a 0 is sensed in that line, the robot moves forward using the routine at the `center` label. It's similar to the previous robot movement routines so I won't go into details. It slowly moves the robot forward by driving both wheels and then jumps back to the `main` loop to test the detectors again.

`LSENS` and `RSENS` are used to correct when the robot is going off the black line. If the "LSENS" detector sees black, the program jumps to the `left` label. At that label, the robot's right wheel is driven to turn the robot back towards the black line. If the "RSENS" detector sees black, then the program jumps to the `right` label. At that label, the robot's left wheel is driven to turn the robot back towards the black line.

These three `IF-THEN` statements do most of the control to make the robot zigzag along the black line. After all the `IF-THEN` statements are complete, the program jumps back to the `main` label using a GOTO command and starts checking the IF-THEN commands all over again.

```
' ----[ Title ]------------------------------   --
'
' File...... proj14pr.BAS
' Purpose... Robot to follow Black Line PIC16F876 -> IRD Line
' Detector
' Author.... Chuck Hellebuyck
' Started... March 1, 2002
' Updated...


' ----[ Program Description ]-------------------   -
'
' This Program uses the 16F876 to control a Servo robot platform.
' The robot has an infrared 3 element line detector module mounted
' under its platform. This program will drive the robot forward
' while it scans for the black line under the detector. If the
' black line hits one of the outer elements, the robot will
' correct back to center as it moves forward. The object is to
' have the robot follow the line from one end to the other.
'
' RA0        GP2D15 Detector Output
' RA1        Left Detector Element Output
' RA2        Center Detector Element Output
' RA3        Right Detector Element Output
' RC1        Left LED
' RC2        Left Center LED
' RC3        Center LED
' RC5        Right Center LED
' RC6        Right LED
' RB2        Right Wheel Servo
' RB5        GP2D15 Servo
' RB7        Left Wheel Servo

' ----[ Revision History ]------------------------
'
'
```

```
' ---[ Constants ]--------------------------------_
'
LSENS var porta.1          'Left detector element
RSENS var porta.2          'Right detector element
CSENS var porta.3          'Center detector element
LLED var portc.1           'Left LED pin
CLED var portc.3           'Center LED pin
RLED var portc.6           'Right LED pin


' ---[ Variables ]--------------------------------_
'
move var byte              'robot movement variable
x var byte                 'general purpose variable



' ---[ Initialization ]---------------------------
'
      define loader_used 1              'Used for bootloader only

Init:
ADCON1 = 7                 'Set portA to digital I/O
TRISA = %11111111          'Set portA to all inputs
TRISC = %00000000          'Set portC to all outputs
TRISB = %00000000          'Set portB to all outputs

' ---[ Main Code ]--------------------------------_
'


main:
if csens = 1 and lsens = 1 and rsens = 1 then error 'Can't find
                                             'line
if lsens = 0 and rsens = 0 then halt    'Both elements found line
if csens = 0 then center   'Center element on line goto center
                           'routine
if lsens = 0 then left     'Left element on line goto left
                           'routine
if rsens = 0 then right    'Right element on line goto right
                           'routine
goto main                  'Loop back and do it again

' *** Center element on line routine
center:
high cled       'Center LED on
low rled        'Right LED off
low lled        'Left LED off
move = 5        'Preset move variable to 5
```

```
gosub fwd            'Jump to fwd subroutine
goto main            'Jump back to main loop

' *** Left element on line routine
left:
high lled            'Left LED on
low cled             'Center LED off
low rled             'Right LED off
move = 5             'Preset move variable to 5
gosub lfwd           'Jump to lfwd subroutine
goto main            'Jump back to main loop

' *** Right element on line routine
right:
high rled            'Right LED on
low cled             'Center LED off
low lled             'Left LED off
move = 5             'Preset move variable to 5
gosub rfwd           'Jump to rfwd subroutine
goto main            'Jump back to main loop

' *** All elements not on line error routine
error:
high rled            'Right LED on
high lled            'Left LED on
high cled            'Center LED on
move = 1             'Preset move variable to 1
gosub fwd            'Jump to fwd subroutine
goto main            'Jump back to main loop

' *** Outer elements sense black line
halt:
high rled            'Right LED on
high lled            'Left LED on
low cled             'Center LED off
pause 10             'Add delay for 10 msec
goto main            'Jump back to main loop

' *** Move robot forward subroutine
FWD:
For x = 1 to move    'Start forward movement loop
pulsout 2, 100       'Turn right wheel forward
pulsout 7, 200       'Turn left wheel forward
pause 10             'Delay to control robot speed
next                 'Repeat forward movement loop
return               'Jump back to where this subroutine was called
```

```
' *** Move robot in reverse subroutine
revrs:
For x = 1 to move    'Start reverse movement loop
pulsout 2, 200       'Turn right wheel in reverse
pulsout 7, 100       'Turn left wheel in reverse
pause 10             'Delay to control robot speed
next                 'Repeat reverse movement loop
return               'Jump back to where this subroutine was called

' *** Move robot forward-left subroutine
Lfwd:
For x = 1 to move    'Start left-forward movement loop
pulsout 2,100        'Turn right wheel forward to go left
pause 10             'Delay to control robot speed
next                 'Repeat left-forward movement loop
return               'Jump back to where this subroutine was called

' *** Move robot forward-right subroutine
Rfwd:
For x = 1 to move    'Start right-forward movement loop
pulsout 7, 200       'Turn left wheel forward to go right
pause 10             'Delay to control robot speed
next                 'Repeat right-forward movement loop
return               'Jump back to where this subroutine was called
```

Final Thoughts

My test track was rather short, so I could test the robot right on my bench. Making a more elaborate test track is a great addition to this project. Make it into a circle with "T" crossings so the robot goes a distance and then stops. Then you can modify the `halt` routine so it delays a period of time and then drives forward. This would make the robot act like a train stopping at the station, but do so automatically.

Another option is to play with the "move" values and try to make the robot move down the line faster. I once saw a robot competition where the object was simply to follow a line. Each team had the same robot and started with the same code. They were allowed to modify the program in any way they wanted to speed up the robot, but it had to follow the line. The winning team was the team that went the full distance of the black line in the fastest time. You could try this by timing your first run and then trying to improve it.

Project #15—Obstacle Detection

A common means of obstacle detection is to bump into something and then back up. That's the method used in the real robot pictured at the beginning of this chapter. I wanted to do something different; a noncontact obstacle detection method was my goal. I decided to use the Sharp GP2D15 infrared detector (IRD) in this project. It can detect objects 10 cm to 80 cm away from its sensor, and outputs a digital high signal when an obstacle is detected. This is easy to monitor with a standard PIC I/O pin. In addition, the Sharp IRD was mounted on top of a servomotor so it could be moved back and forth in a sweeping motion to look for obstacles all across the front of the robot (sort of a "roving eye").

To test my idea out, I wanted to see if the robot could find the only opening in a large enclosed area without bumping into the walls. For the most part it worked, but more than one IRD would make it better at detecting obstacles; the single IRD has a limited range.

Because the IRD cannot detect anything less than 10 cm, the servomotor/IRD assembly was positioned 10 cm back from the front of the robot. This allowed the robot to see objects right in front without a dead zone. As a final step, three LEDs were used to show the detected direction. One for each detectable position was used: left, center, and right. This helped to show what the robot was "thinking" and also helped debug the software.

I made the arena out of 8.5x11-inch sheets of printer paper taped together. I folded the edge to make the base. I wanted it to be weak so the robot would not be able to bump into it and keep going. Instead, this method allowed the robot to drive right over the walls if something did not work. When it was finished, the robot found the opening every time. Figures 9-8 through 9-12 show the completed robot, with servomotor IRD "eye," in action. The pictures show the robot as it finds the opening and begins to drive through it.

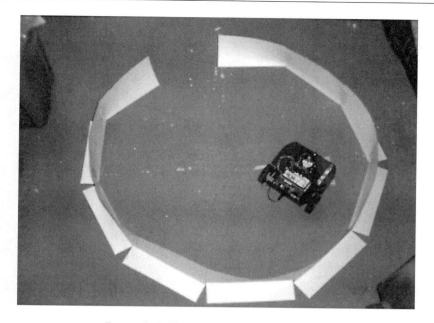

Figure 9-8: The robot inside the "arena."

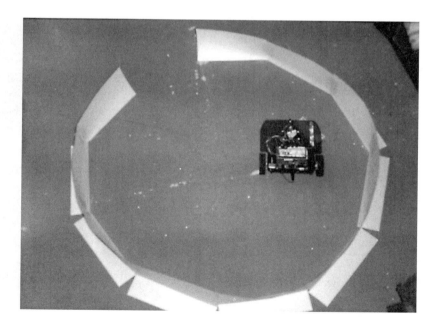

Figure 9-9: The robot looks for an opening.

Figure 9-10: The robot moves toward the opening.

Figure 9-11: The robot enters the opening of the "arena."

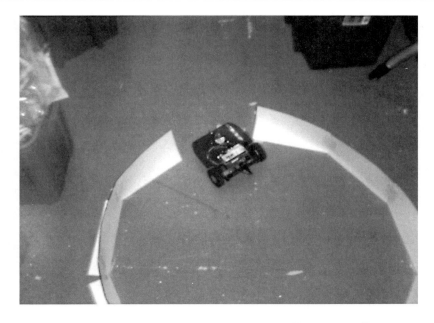

Figure 9-12: Success! The robot "escapes" the "arena."

Hardware

The schematic in Figure 9-13 shows the circuit for this project. The same connections are used for the drive servomotors on Port B pin 2 and pin 7. The third servomotor for the IRD is connected to Port B pin 5.

The IRD is connected to Port A pin 0. A single 10k pull-up resistor is attached to the sense lead as required by the Sharp GP2D15 data sheet. The direction LEDs are driven by Port C. By putting all the LEDs on a common port, they can be shut off or turned on with a single command.

The IRD servomotor is different from the drive servomotors. The IRD servomotor is not reworked for continuous rotation. It is a standard servomotor that drives back and forth based on the pulse of the signal it is sent. The rest of the connections are the same as in all previous projects.

I attached the Sharp sensor to the servomotor with double-sided sticky foam tape. It worked well, but an even stronger mounting would have been better. The wires for the sensor were a bit stiff, and the constant back and forth motion caused the wires to push the sensor. After time, the sensor would lean forward and cause strange readings. If you build this, make a better mount or use a more flexible wire than the ones that are typically available for the sensor.

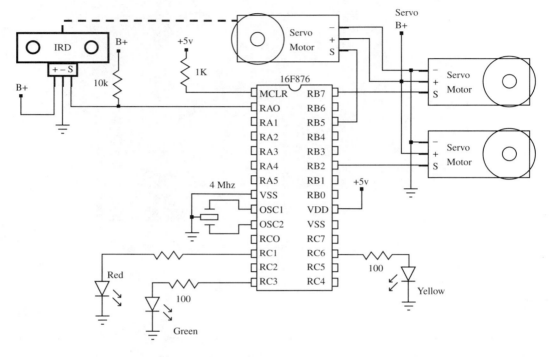

Figure 9-13: Circuit for the robot's IRD sensor and servomotor circuitry.

PBC Code

The code has similarities to the two previous projects. It uses the same subroutines as before. The biggest difference is the code to read and control the sensor. It was trickier than I thought to drive the robot and the sensor servomotor without having very jumpy movement. I learned a little more about controlling a servomotor, and as a result I came to understand one of the reasons why larger, more complex robots use geared DC motors and drive them directly through separate control electronics. It gives the programmer more control than relying on the internal servomotor electronics.

The first step in this code is setting up the constants and variables:

```
' ——-[ Constants ]——————————————————————-
'
symbol GP2D15 = bit0        'IRD sensor connected to PortA bit0
symbol adcon1 = $9F         'A/D control register location
symbol trisa = $85          'PortA Tris register location
symbol trisc = $87          'PortC Tris register location
symbol porta = 5            'PortA register location
symbol portc = 7            'PortC register location


' ——-[ Variables ]——————————————————————-
'
symbol position = b2        'obstacle position variable
symbol move = b3            'robot movement variable
symbol x = b4               'general purpose variable
symbol scan = b5            'GP2D15 position variable
```

The GP2D15 name is tied to bit0 of the "B0" variable. This way we can PEEK PortA and then just read bit 0 to know the state of the GP2D15.

Next we make constant names for all the important registers: ADCON1, TRISA, TRISC, PORTA and PORTC. This makes the code easier to read and the PEEK and POKE command lines easier to write. After the constants are set, the variables are established. We use the same "move" and "x" variables as before and add two more. The first is "position," which indicates where the sensor is pointing and the second is "scan," which controls the movement of the sensor's servomotor.

The init label begins our command code:

```
Init:
poke ADCON1,7               'Set portA to digital I/O
poke TRISA, %11111111       'Set portA to all inputs
poke TRISC, %00000000       'Set portC to all outputs
DIRS = %11111111            'Set portB to all outputs
scan = 100                  'Preset GP2D15 variable to 100
```

First we POKE the ADCON1 register to 7, which makes Port A all digital I/O. Next we POKE TRISA and TRISC to all inputs and all outputs, respectively. Port B is controlled with the DIRS directive, so we make Port B all outputs by making the DIRS directives all ones. Finally we preset the "scan" variable to 100.

The `main` code loop is entered next. It's much longer than the previous `main` loops so I'm going to break it up a bit here at the beginning, which is the scan routine. Later on, the loop performs functions based on whether an obstacle was or was not detected in the scan routine.

First, we POKE Port C with 0 to turn off all LEDs. Then we preset the "position" variable to 0. You may be wondering why I didn't just do that at the `init` label. I did this because I wanted them refreshed every time the program went through the `main` loop.

```
Main:
poke portc, 0            'Set all LEDs off
position = 0             'Preset position to 0
```

The next section is the `scan` routine. A FOR-NEXT loop is used to step the servomotor through 17 steps of movement from one side of the robot to the other. Normally a servomotor only works from 100 msec to 200 msec. I found the servomotor was not reacting quickly enough, but when I went outside those boundaries it began to work so I kept it in.

```
' *** Scan routine
for scan = 70 to 230 step 10    'Start scanning GP2D15 right to
                                'left
pulsout 5, scan                 'Send scanning position signal to
                                'servo
pause 20                        'Delay fo Servo to react
peek porta, b0                  'Read porta
if GP2D15 = 0 then skp          'If obstacle not detected, jump to
                                'skp
if scan <= 150 then pas         'Test value of scan
position = 1                    'Scan less or equal to 150, set position
                                'to 1
goto skp                        'Bypass scan > 150 routine
pas:
position = 2                    'Scan greater than 150, set position to
                                '2
skp:
next                            'Loop back for next GP2D15
                                'position
```

The program sends the servomotor pulse out using the PULSOUT command. Then a 20-msec delay is added to allow the servomotor to react. The program quickly scans the GP2D15 bit to see if it is 0, which indicates no obstacle was detected. A 1 indicates an obstacle was detected. If an obstacle is not detected, the program jumps to the skp label and then jumps back to the next value of the FOR-NEXT loop.

If an object was detected, the program then tests the "scan" variable value. If the value is less than or equal to 150, then the "position" variable is set to 1. If the value is greater than 150, then the "position" variable is set to 2. Later in the program we read that variable to determine if the obstacle was detected to the right and/or center of the robot or to the left of the robot. We use that to decide which way to turn.

We do this until the servomotor has completely scanned from right to left. Then we take the last value of the "position" variable and jump to the obstacle detected routine:

```
' *** Obstacle Detected

if position <> 0 then decide       'Test for obstacle in previous
                                   'loop
                                   ' if detected jump to decide
                                   ' routine
```

First we test the "position" variable to see if it is 0. If it is, then we continue on to the obstacle not detected routine. If it is not 0, then we jump to the decide label where we decide how to drive the robot away from the obstacle.

Let's first cover the obstacle not detected routine:

```
' *** Obstacle not detected, Forward movement routine
poke portc,0                       'All LEDs off
poke portc, %00001000              'Green LED on
move = 6                           'Preset move variable for forward
                                   'movement
for x = 1 to move                  'Start loop for servo wheel
                                   'control
pulsout 2, 100                     'Turn Right wheel servo
pulsout 7, 200                     'Turn left wheel servo
pause 10                           'Delay for 10 msec to slow down
                                   'robot
next                               'Repeat the move robot routine
Ret:
goto main                          'Jump back to main label and start
                                   'over
```

The first step is to turn all the LEDs off and then turn just the center green LED on. This indicates the robot "believes" it's free to move forward. We do this similarly to the previous projects by setting the "move" variable to a value. The difference is we don't call the FWD subroutine that we did in the other projects. I put the FWD subroutine code right in the main loop so the robot will drive forward without delay. (I've covered that code already in the previous projects, so I won't bore you with it again.) After the robot's "move forward" routine is complete, the program jumps back to the main label to scan again for obstacles.

Now if an obstacle is detected, the program jumps to the decide label. That code is:

```
decide:
branch position,(main,right,left) 'Choose correction routine
goto main                          'Jump to main routine if position
                                   '> 2
```

As you can see, this section is very short and just uses a BRANCH command to redirect the code. The BRANCH command has three options. If the "position" variable is 0, then the program jumps back to the main label. This should never occur since the program won't jump to decide if "position" is zero. If "position" is 1, then the program jumps to the right label, where the program will react to an obstacle on the right or center of the robot. If "position" is 2, then the program jumps to the left label to react to an obstacle on the left of the robot.

Both of the routines are similar to the previous projects in this chapter because they move one wheel to turn the robot either right or left. In the previous project, we did it to find the line. In this project, we do it to move away from the paper wall. Once the robot corrects for the obstacle, it jumps back to the main loop and starts the scanning all over again. The robot usually found its way out rather quickly!

```
' ---[ Title ]------------------------------
'
' File...... proj15pb.BAS
' Purpose... Robot with obstacle detection PIC16F876 -> Sharp
' GP2D15
' Author.... Chuck Hellebuyck
' Started... March 20, 2002
' Updated...
```

```
' —-[ Program Description ]————————————--
'
' This Program uses the 16F876 to control a Servo robot platform.
' The robot has a servo mounted at its front with a Sharp GP2D15
' infrared detector on top. This program will drive the robot
' forward while it scans for obstacles in front of its path. If an
' obstacle is detected, the robot will turn away and then move
' forward. The object is to have the robot find its way out of a
' boxed in arena with only one opening for the exit.
'
' RA0         GP2D15 output
' RC1         Left LED (Red)
' RC3         Center LED (green)
' RC6         Right LED (yellow)
' RB2         Right Wheel Servo
' RB5         GP2D15 Servo
' RB7         Left Wheel Servo

' —-[ Revision History ]————————————
'
'

' —-[ Constants ]————————————————-
'
symbol GP2D15 = bit0        'IRD sensor connected to PortA bit0
symbol adcon1 = $9F         'A/D control register location
symbol trisa = $85          'PortA Tris register location
symbol trisc = $87          'PortC Tris register location
symbol porta = 5            'PortA register location
symbol portc = 7            'PortC register location

' —-[ Variables ]————————————————-
'
symbol position = b2        'obstacle position variable
symbol move = b3            'robot movement variable
symbol x = b4               'general purpose variable
symbol scan = b5            'GP2D15 position variable

' —-[ Initialization ]————————————————
```

```
'
Init:
poke ADCON1,7              'Set portA to digital I/O
poke TRISA, %11111111      'Set portA to all inputs
poke TRISC, %00000000      'Set portC to all outputs
DIRS = %11111111           'Set portB to all outputs
scan = 100                 'Preset GP2D15 variable to 100


' —-[ Main Code ]————————————————————-
'

Main:
poke portc, 0              'Set all LEDs off
position = 0               'Preset position to 0

' *** Scan routine
for scan = 70 to 230 step 10   'Start scanning GP2D15 right to
                               'left

pulsout 5, scan                'Send scanning position signal to
                               'servo

pause 20                       'Delay for Servo to react
peek porta, b0                 'Read porta
if GP2D15 = 0 then skp         'If obstacle not detected, jump to
                               'skp

if scan <= 150 then pas        'Test value of scan
position = 1                   'Scan less or equal to 150, set position
                               'to 1

goto skp                       'Bypass scan > 150 routine
pas:
position = 2                   'Scan greater than 150, set position to
                               '2

skp:
next                           'Loop back for next GP2D15
                               'position


' *** Obstacle Detected

if position <> 0 then decide   'Test for obstacle in previous
                               'loop
                               ' if detected jump to decide
                               ' routine


' *** Obstacle not detected, Forward movement routine
poke portc,0                   'All LEDs off
poke portc, %00001000          'Green LED on
move = 6                       'Preset move variable for forward
                               'movement

for x = 1 to move              'Start loop for servo wheel
                               'control
```

```
pulsout 2, 100              'Turn Right wheel servo
pulsout 7, 200              'Turn left wheel servo
pause 10                    'Delay for 10 msec to slow down
                            'robot
next                        'Repeat the move robot routine
Ret:
goto main                   'Jump back to main label and start
                            'over


decide:
branch position,(main,right,left) 'Choose correction routine
goto main                   'Jump to main routine if position
                            '> 2


' *** Left obstacle detected, move right
left:
poke portc, 0               'Turn off all LEDs
poke portc, %00000010       'Turn on Left LED
move = 6                    'Preset move variable to 6
for x = 1 to move           'Start turn right routine
pulsout 7, 200              'Turn left wheel servo
pause 30                    'Delay to control robot speed
next                        'Repeat the move robot routine
goto ret                    'Jump back to main loop at Ret label


' *** Right obstacle detected, move left
right:
poke portc,0                'Turn off all LEDs
poke portc, %01000000       'Turn on Right LED
move = 6                    'Preset move variable to 6
for x = 1 to move           'Start turn left routine
pulsout 2, 100              'Turn right wheel servo
pause 30                    'Delay to control robot speed
next                        'Repeat the move robot routine
goto ret                    'Jump back to main loop at Ret label


goto main                   'End program by jumping to main
```

PBPro Code

As in the PBC example, the PBPro code has similarities to that used in the two previous projects (for example, it uses the same subroutines as before). The biggest difference is the code to read and control the sensor. It was trickier than I thought to smoothly drive the robot and the sensor servomotor.

The first step is to set the constants and variables:

```
' —-[ Constants ]————————————--
'
LLED var portc.1          'Left LED pin
CLED var portc.3          'Center LED pin
RLED var portc.6          'Right LED pin


' —-[ Variables ]————————————--
'
position var byte          'obstacle position variable
move var byte              'robot movement variable
x var byte                 'general purpose variable
scan var byte              'GP2D15 position variable
```

We simplify writing our code by giving LED names to the Port C pins they are connected to, using the VAR directive. After the constants are set, the variables are established. We use the same "move" and "x" variables as before, and add two more: "position" to indicate where the sensor is pointing, and "scan" to control the movement of the sensor's servomotor.

The init label begins our command code after we establish the bootloader DEFINE:

```
define loader_used 1              'Used for bootloader only

Init:
ADCON1 = 7                        'Set portA to digital I/O
TRISA = %11111111                 'Set portA to all inputs
TRISC = %00000000                 'Set portC to all outputs
TRISB = %00000000                 'Set portB to all outputs
scan = 100                        'Preset GP2D15 variable to 100
```

First we set the ADCON1 register to 7, which makes Port A all digital I/O. Next we set Port A to inputs and Port B and Port C to all outputs. Finally, we preset the "scan" variable to 100.

The main code loop is entered next. It's much longer than the main loops in the previous projects in this chapter, so I'm going to break it up a bit here. At the beginning is the scan routine. Later the loop performs functions based on whether an obstacle was detected or not in the scan routine.

First we preset Port C with 0 to turn off all LEDs. Then we preset the "position" variable to 0. Why didn't I just do that at the `init` label? The reason is because I wanted them refreshed every time the program went through the `main` loop.

```
Main:
portc = 0                       'Set all LEDs off
position = 0                    'Preset position to 0
```

The next section is the `scan` routine. A FOR-NEXT loop is used to step the servomotor through 17 steps of movement from one side of the robot to the other. Normally a servomotor only works from 100 msec to 200 msec, but I found the servo was not reacting quickly enough. But when I went outside those boundaries, it began to work, so I kept it in.

```
' *** Scan routine
for scan = 70 to 230 step 10    'Start scanning GP2D15 right to
                                'left
pulsout 5, scan                 'Send scanning position signal to
                                'servo
pause 20                        'Delay fo Servo to react
if porta.0 = 0 then skp         'If obstacle not detected, jump to
                                'skp
if scan <= 150 then pass        'Test value of scan
position = 1                  'Scan less or equal to 150, set position
                              'to 1
goto skp                        'Bypass scan > 150 routine
pass:
position = 2                  'Scan greater than 150, set position to
                              '2
skp:
next                            'Loop back for next GP2D15
                                'position
```

The program sends the servomotor pulse out using the PULSOUT command. Next a 20-msec delay is added to give the servomotor time to react. The program quickly scans the Port A pin 0 connected to the GP2D15 to see if it's 0. A 0 indicates no obstacle was detected while a 1 indicates an obstacle was detected. If an obstacle is not detected, the program jumps to the `skp` label and then jumps back to the next value of the FOR-NEXT loop.

If an object was detected, the program then tests the "scan" variable value. If the value is less than or equal to 150, then the "position" variable is set to 1. If the value is greater than 150, then the "position" variable is set to 2. Later in the program we read that variable to determine if the obstacle was detected to the right and/or center of the robot or to the left of the robot. We use that to decide which way to turn. We do this until the servomotor has completely scanned from right to left. Then we take the last value of the "position" variable and jump to the obstacle detected routine below:

```
' *** Obstacle Detected

if position <> 0 then decide      'Test for obstacle in previous
                                  'loop
                                  ' if detected jump to decide
                                  ' routine
```

First we test the "position" variable to see if it is 0. If it is, then we continue on to the obstacle not detected routine. If it is not 0, then we jump to the decide label where we decide how to drive the robot away from the obstacle.

Let's first examine the obstacle not detected routine:

```
' *** Obstacle not detected, Forward movement routine
portc = 0                         'All LEDs off
high cled                         'Green LED on
move = 6                          'Preset move variable for forward
                                  'movement
for x = 1 to move                 'Start loop for servo wheel
                                  'control
pulsout 2, 100                    'Turn Right wheel servo
pulsout 7, 200                    'Turn left wheel servo
pause 10                          'Delay for 10 msec to slow down
                                  'robot
next                              'Repeat the move robot routine
Ret:
goto main
```

The first step is to turn all the LEDs off and then turn just the center green LED on. This indicates the robot "believes" it's free to move forward. We do this similarly to the previous projects, by setting the "move" variable to a value. The difference is we don't call the FWD subroutine as we did in earlier projects because I put the FWD subroutine code in the main loop so the robot will drive forward without delay. Since I've covered that code in the previous projects, I won't repeat it here.

After the robot's "move forward" routine is complete, the program jumps back to the `main` label to scan again for obstacles. If an obstacle was detected, the program would jump to the `decide` label. That code is below.

```
decide:
branch position,[main,right,left] 'Choose correction routine
goto main                         'Jump to main routine if position
                                  '> 2
```

This section is short and uses a BRANCH command to redirect the code. The BRANCH command has three options. If the "position" variable is 0, then the program jumps back to the `main` label. (This should never occur since the program won't jump to `decide` if "position" is zero.) If "position" is 1, then the program jumps to the `right` label where the program will react to an obstacle on the right or center of the robot. If "position" is 2, then the program jumps to the `left` label to react to an obstacle on the left of the robot.

Both of the routines are similar to those in previous projects because they move one wheel to turn the robot either right or left. In the previous project, we did it to find the line; in this project, we do it to move away from the paper wall. Once the robot corrects for the obstacle, it jumps back to the `main` loop and starts the scanning all over again.

```
' —-[ Title ]———————————————————--
'
' File...... proj15pr.BAS
' Purpose... Robot with obstacle detection PIC16F876 -> Sharp
' GP2D15
' Author.... Chuck Hellebuyck
' Started... March 20, 2002
' Updated...


' —-[ Program Description ]————————————
'
' This Program uses the 16F876 to control a Servo robot platform.
' The robot has a servo mounted at its front with a Sharp GP2D15
' infrared detector on top. This program will drive the robot
' forward while it scans for obstacles in front of its path. If an
' obstacle is detected, the robot will turn away and then move
' forward. The object is to have the robot find its way out of a
' boxed in arena with only one opening for the exit.
```

```
'
' RA0          GP2D15 output
' RC1          Left LED
' RC2          Left Center LED
' RC3          Center LED
' RC5          Right Center LED
' RC6          Right LED
' RB2          Right Wheel Servo
' RB5          GP2D15 Servo
' RB7          Left Wheel Servo


' ----[ Revision History ]-----------------------
'
'


' ----[ Constants ]------------------------------
'
LLED var portc.1          'Left LED pin
CLED var portc.3          'Center LED pin
RLED var portc.6          'Right LED pin

' ----[ Variables ]-----------------------------
'
position var byte         'obstacle position variable
move var byte             'robot movement variable
x var byte                'general purpose variable
scan var byte             'GP2D15 position variable

' ----[ Initialization ]-------------------------
'
        define loader_used 1            'Used for bootloader only

Init:
ADCON1 = 7                'Set portA to digital I/O
TRISA = %11111111         'Set portA to all inputs
TRISC = %00000000         'Set portC to all outputs
TRISB = %00000000         'Set portB to all outputs
scan = 100                'Preset GP2D15 variable to 100

' ----[ Main Code ]------------------------------
'
Main:
portc = 0                 'Set all LEDs off
position = 0              'Preset position to 0
```

```
' *** Scan routine
for scan = 70 to 230 step 10      'Start scanning GP2D15 right to
                                  'left
pulsout 5, scan                   'Send scanning position signal to
                                  'servo
pause 20                          'Delay fo Servo to react
if porta.0 = 0 then skp           'If obstacle not detected, jump to
                                  'skp
if scan <= 150 then pass          'Test value of scan
position = 1                 'Scan less or equal to 150, set position
                             'to 1
goto skp                          'Bypass scan > 150 routine
pass:
position = 2                 'Scan greater than 150, set position to
                             '2
skp:
next                              'Loop back for next GP2D15
                                  'position


' *** Obstacle Detected

if position <> 0 then decide      'Test for obstacle in previous
                                  'loop
                                  ' if detected jump to decide
                                  ' routine


' *** Obstacle not detected, Forward movement routine
portc = 0                         'All LEDs off
high cled                         'Green LED on
move = 6                          'Preset move variable for forward
                                  'movement
for x = 1 to move                 'Start loop for servo wheel
                                  'control
pulsout 2, 100                    'Turn Right wheel servo
pulsout 7, 200                    'Turn left wheel servo
pause 10                          'Delay for 10 msec to slow down
                                  'robot
next                              'Repeat the move robot routine
Ret:
goto main                         'Jump back to main label and start
                                  'over
```

```
decide:
branch position,[main,right,left] 'Choose correction routine
goto main                         'Jump to main routine if position
                                  '> 2


' *** Left obstacle detected, move right
left:
portc = 0                'Turn off all LEDs
high lled                'Turn on Left LED
move = 6                 'Preset move variable to 6
for x = 1 to move        'Start turn right routine
pulsout 7, 200           'Turn left wheel servo
pause 30                 'Delay to control robot speed
next                     'Repeat the move robot routine
goto ret                 'Jump back to main loop at Ret label

' *** Right obstacle detected, move left
right:
portc = 0                'Turn off all LEDs
high rled                'Turn on Right LED
move = 6                 'Preset move variable to 6
for x = 1 to move        'Start turn left routine
pulsout 2, 100           'Turn right wheel servo
pause 30                 'Delay to control robot speed
next                     'Repeat the move robot routine
goto ret                 'Jump back to main loop at Ret label


goto main                'End program by jumping to main
```

Final Thoughts

My young daughter thought it was really neat that the robot "knew" where the opening was. Occasionally the robot would miss the opening the first time around. It would just loop around again and usually find it the second trip. When that would happen, she would talk to the robot and tell it to go back because it had missed the opening. She was only five, but it just proved to me that this is a great project to show children how robots work.

One thing I thought about doing was to make a charging station or download station for the robot. The robot could be reworked to collect data with a digital camera or sensors and then find the docking station using an approach similar to this project. It will be more complex than this project, but should be easy to build it up using this program as the starting point.

In Conclusion. . .

This chapter concludes this book with the exception of the appendices. This last project represents how each project evolved. Each one was filled with walls, but eventually I found the opening and moved on to the next project!

I hope you likewise find the openings in your "walls" as you work with PICs and PicBasic.

APPENDIX A
PicBasic and Project Resources

Here is a list of resources used to develop the projects and code in this book, and I recommend that you check them out. Through these sources you can find the many tools and components I used to make the projects in this book. I continue to use these sites as access for information regarding the ever-expanding world of embedded Basic programming.

Acroname
Offers robotics accessories of various types, including the Sharp sensor.

> www.acroname.com
> 4894 Sterling Dr.
> Boulder CO, 80301
> (720) 564-0373
> email: sales@acroname.com

Basic Micro
Offers MBasic compilers, development boards, programmers, Atom module, and robotic accessories.

> www.basicmicro.com
> 22882 Orchard Lake Rd.
> Farmington Hills, MI. 48336
> (248) 427-0040
> (734) 425-1722 fax
> email: sales@basicmicro.com

Chuck Hellebuyck's Electronic Products
My web site! I Offer PicBasic, Atom modules, programmers, development boards, and robotic accessories, including reworked servomotors. I'll also be posting updates to this book, including this resource list, at the URL below.

> www.elproducts.com
> 1775 Medler
> Commerce, MI 48382
> (248) 515-4264
> email: chuck@elproducts.com

Lynxmotion Inc.
Offers robotics kits and accessories including the robot base used in Chapter 9.

> www.lynxmotion.com
> PO Box 818
> Pekin, IL 61555-0818
> 866-512-1024
> 309-382-1254 fax
> sales@lynxmotion.com

microEngineering Labs, Inc.
Offers: PicBasic compilers, development boards, programmers, and bootloader.

> www.melabs.com
> Box 60039
> Colorado Springs CO 80960
> (719) 520-5323
> (719) 520-1867 fax
> email: support@melabs.com

Parallax

Offers Basic Stamp modules and robotics accessories.

599 Menlo Drive
Suite 100
Rocklin, California 95765
(888) 512-1024
(916) 624-8003
email: sales@parallaxinc.com

Reworking servomotors

The hardest part about building a robot is reworking the servomotors. As we discussed in the previous chapter, servomotors are designed to drive back and forth based on a pulse-width modulated signal. In order to use these types of motors for a robot drive, the internals have to be reworked to spin a full 360 degrees in both directions. A great explanation of how to rework a servomotor can be found at this Internet link:

http://www.acroname.com/robotics/info/ideas/continuous/continuous.html

APPENDIX B
ASCII Table

Decimal	Octal	Hex	Binary	Value	
000	000	000	00000000	NUL	(Null char.)
001	001	001	00000001	SOH	(Start of Header)
002	002	002	00000010	STX	(Start of Text)
003	003	003	00000011	ETX	(End of Text)
004	004	004	00000100	EOT	(End of Transmission)
005	005	005	00000101	ENQ	(Enquiry)
006	006	006	00000110	ACK	(Acknowledgment)
007	007	007	00000111	BEL	(Bell)
008	010	008	00001000	BS	(Backspace)
009	011	009	00001001	HT	(Horizontal Tab)
010	012	00A	00001010	LF	(Line Feed)
011	013	00B	00001011	VT	(Vertical Tab)
012	014	00C	00001100	FF	(Form Feed)
013	015	00D	00001101	CR	(Carriage Return)
014	016	00E	00001110	SO	(Shift Out)
015	017	00F	00001111	SI	(Shift In)
016	020	010	00010000	DLE	(Data Link Escape)
017	021	011	00010001	DC1 (XON)	(Device Control 1)
018	022	012	00010010	DC2	(Device Control 2)
019	023	013	00010011	DC3 (XOFF)	(Device Control 3)

Decimal	Octal	Hex	Binary	Value	
020	024	014	00010100	DC4	(Device Control 4)
021	025	015	00010101	NAK	(Negative Acknowledgement)
022	026	016	00010110	SYN	(Synchronous Idle)
023	027	017	00010111	ETB	(End of Trans. Block)
024	030	018	00011000	CAN	(Cancel)
025	031	019	00011001	EM	(End of Medium)
026	032	01A	00011010	SUB	(Substitute)
027	033	01B	00011011	ESC	(Escape)
028	034	01C	00011100	FS	(File Separator)
029	035	01D	00011101	GS	(Group Separator)
030	036	01E	00011110	RS	(Request to Send)(Record Separator)
031	037	01F	00011111	US	(Unit Separator)
032	040	020	00100000	SP	(Space)
033	041	021	00100001	!	
034	042	022	00100010	"	
035	043	023	00100011	#	
036	044	024	00100100	$	
037	045	025	00100101	%	
038	046	026	00100110	&	
039	047	027	00100111	'	
040	050	028	00101000	(	
041	051	029	00101001	)	
042	052	02A	00101010	*	
043	053	02B	00101011	+	
044	054	02C	00101100	,	
045	055	02D	00101101	-	
046	056	02E	00101110	.	

Decimal	Octal	Hex	Binary	Value
047	057	02F	00101111	/
048	060	030	00110000	0
049	061	031	00110001	1
050	062	032	00110010	2
051	063	033	00110011	3
052	064	034	00110100	4
053	065	035	00110101	5
054	066	036	00110110	6
055	067	037	00110111	7
056	070	038	00111000	8
057	071	039	00111001	9
058	072	03A	00111010	:
059	073	03B	00111011	;
060	074	03C	00111100	<
061	075	03D	00111101	=
062	076	03E	00111110	>
063	077	03F	00111111	?
064	100	040	01000000	@
065	101	041	01000001	A
066	102	042	01000010	B
067	103	043	01000011	C
068	104	044	01000100	D
069	105	045	01000101	E
070	106	046	01000110	F
071	107	047	01000111	G
072	110	048	01001000	H
073	111	049	01001001	I
074	112	04A	01001010	J
075	113	04B	01001011	K

Decimal	Octal	Hex	Binary	Value
076	114	04C	01001100	L
077	115	04D	01001101	M
078	116	04E	01001110	N
079	117	04F	01001111	O
080	120	050	01010000	P
081	121	051	01010001	Q
082	122	052	01010010	R
083	123	053	01010011	S
084	124	054	01010100	T
085	125	055	01010101	U
086	126	056	01010110	V
087	127	057	01010111	W
088	130	058	01011000	X
089	131	059	01011001	Y
090	132	05A	01011010	Z
091	133	05B	01011011	[
092	134	05C	01011100	\
093	135	05D	01011101	]
094	136	05E	01011110	^
095	137	05F	01011111	_
096	140	060	01100000	`
097	141	061	01100001	a
098	142	062	01100010	b
099	143	063	01100011	c
100	144	064	01100100	d
101	145	065	01100101	e
102	146	066	01100110	f
103	147	067	01100111	g
104	150	068	01101000	h

Decimal	Octal	Hex	Binary	Value
105	151	069	01101001	i
106	152	06A	01101010	j
107	153	06B	01101011	k
108	154	06C	01101100	l
109	155	06D	01101101	m
110	156	06E	01101110	n
111	157	06F	01101111	o
112	160	070	01110000	p
113	161	071	01110001	q
114	162	072	01110010	r
115	163	073	01110011	s
116	164	074	01110100	t
117	165	075	01110101	u
118	166	076	01110110	v
119	167	077	01110111	w
120	170	078	01111000	x
121	171	079	01111001	y
122	172	07A	01111010	z
123	173	07B	01111011	{
124	174	07C	01111100		
125	175	07D	01111101	}
126	176	07E	01111110	~
127	177	07F	01111111	DEL

Index